RAND PROJECT AIR FORCE

Gaming Space

A Game-Theoretic Methodology for Assessing the Deterrent Value of Space Control Options

Forrest E. Morgan, Gary McLeod, Michael Nixon, Christopher Lynch, Myron Hura

Prepared for the United States Air Force

Approved for public release; distribution unlimited

For more information on this publication, visit www.rand.org/t/RR694

Library of Congress Cataloging-in-Publication Data is available for this publication.
ISBN: 978-0-8330-9790-3

Published by the RAND Corporation, Santa Monica, Calif.

Preface

The Chinese direct-ascent anti-satellite weapon test in January 2007 was a stark reminder that potential adversaries are pursuing offensive space control (OSC) systems and may seek to limit U.S. freedom of action in space. Adversaries have already employed non-kinetic OSC capabilities, such as Global Positioning System jammers, in recent conflicts, and they might attempt even more provocative attacks in the future, in efforts to disrupt space operations and decrease the benefits the joint warfighter derives from U.S. space systems.[1] The capabilities needed to attack U.S. space systems are becoming less expensive and proliferating on the world market. Left unchallenged, such developments increase the chances that attacks on U.S. space systems will become more common.

The U.S. National Space Policy supports activities, including deterrence, that reduce the likelihood of such attacks on U.S. space systems. Strategies for deterrence attempt to alter the adversary's cost-benefit calculus to make attacks on U.S. space systems seem unattractive by convincing the adversary that an attack would be ineffective or would result in serious consequences. The U.S. Air Force can play a key role in supporting deterrence strategies by developing and operating space systems that can demonstrate U.S. capability and intent to potential adversaries. For example, defensive space control (DSC) demonstrations may be effective in convincing a potential adversary that its attacks could be easily nullified. As a result, the adversary may be deterred from employing its OSC systems during crisis or conflict, or, better yet, be dissuaded from developing such systems in the first place.

However, deciding which defensive systems to develop is a complex problem, because multiple approaches are conceivable for making the U.S. space infrastructure more resilient and defending the services it provides. Some DSC options may contribute more to deterrence than others. Some options might be initially effective but easily countered, or they may escalate the conflict in ways that are ultimately costly to U.S. interests. And some approaches for defending U.S. space capabilities could be viewed as dangerous or provocative, antagonizing to third parties, and generating political costs for

---

[1] Donna Miles, "Iraq Jamming Incident Underscores Lessons About Space," *American Forces Press Service*, September 15, 2004; Lt Gen Dan P. Leaf, "Providing Combat Effects to the Battlefield," *High Frontier*, Vol. 1, No. 4, 2005, pp. 4–5; LTG Larry J. Dodgen, "Leveraging Space to Support the Changing Paradigm," *High Frontier*, Vol. 1, No. 4, 2005, pp. 6–11.

the United States. This report provides an analytical framework and describes a tool RAND Project AIR FORCE has developed to help decisionmakers organize and weigh these concerns in a structured, objective manner.

This report should be of interest to U.S. Department of Defense (DoD) personnel developing capabilities, plans, and options to deter, defend against, and, if necessary, defeat efforts to interfere with or attack U.S. or allied space systems. It should also be of interest to other U.S. government personnel responsible for developing and implementing national and DoD space policies and strategies.

The research reported here was sponsored by the executive director of the Space and Missile Systems Center, and was conducted within the Force Modernization and Employment Program of RAND Project AIR FORCE as part of the project "Assessing the Deterrent Value of Defensive Space Control Options." The purpose of the project was to develop a methodology to assess the extent to which DSC approaches are consistent with space deterrence strategies.

This report leverages and extends prior Project AIR FORCE research, which is documented in Forrest E. Morgan, *Deterrence and First-Strike Stability in Space: A Preliminary Assessment*, MG-916-AF, 2010. This monograph argues that first-strike stability in space appears to be eroding and that the United States should take concerted action to strengthen that stability by developing a strategy to deter future adversaries from attacking U.S. space systems. Space stability is a fundamental U.S. national security interest. War in space would likely be costly for the United States, even if it were to "win" such a conflict and achieve dominance of that domain. Therefore, U.S. space policies and strategies would better serve the public interest if they were explicitly crafted to deter such conflicts while retaining capabilities to win them in the event of deterrence failure.

RAND Project AIR FORCE

RAND Project AIR FORCE (PAF), a division of the RAND Corporation, is the U.S. Air Force's federally funded research and development center for studies and analyses. PAF provides the Air Force with independent analyses of policy alternatives affecting the development, employment, combat readiness, and support of current and future air, space, and cyber forces. Research is conducted in four programs: Force Modernization and Employment; Manpower, Personnel, and Training; Resource Management; and Strategy and Doctrine. The research reported here was prepared under contract FA7014-06-C-0001.

Additional information about PAF is available on our website: www.rand.org/paf

This report documents work originally shared with the U.S. Air Force in November 2013. The draft report, issued in April 2014, was reviewed by formal peer reviewers and U.S. Air Force subject-matter experts.

Contents

Figures

Tables

Summary

U.S. space capabilities are at risk. Potential adversaries are well aware that U.S. transformational warfighting capabilities are increasingly reliant on support from space systems. As a result, they may be developing offensive space control (OSC) weapons to hold U.S. space systems at risk in crisis or attack them in war. Consistent with the inherent right of self-defense, the United States has stated it will defend its space systems.

The objective of the research in this report is to help the Air Force assess the deterrent value of alternative defensive space control (DSC) options.[1] Specifically, we sought to develop a methodology to identify those DSC options that would likely contribute to deterrence and those DSC options that would likely generate escalation dynamics or political costs that could further imperil U.S. interests.

The research involved a four-step process. First, we reviewed national-level and U.S. Department of Defense (DoD)-level space policy and strategy documents for DSC guidance and direction and stance on deterrence. Next, we identified a range of OSC options that could threaten space force enhancement systems. We then identified a range of DSC options for defending against each of those OSC threats. Finally, we developed a methodology, based on sequential game theory, to assess the deterrent value, escalation dynamics, and potential political costs of the various DSC options.

Explaining the Methodology

This report introduces and explains a game-theoretic methodology for assessing the potential effects of alternative approaches to space control. Game theory is a branch of social science that applies structured, logical approaches in the analysis of interactive decisionmaking. It is used to identify the most-effective strategies possible in contests between rational, intelligent opponents. Multiple branches of game theory have been developed since it was first applied to national defense problems in the early Cold War. The branch of game theory most helpful for identifying optimal strategies for DSC development is sequential game theory, in which each opponent moves in turn,

[1] To ensure the analysis captures all possible ways to make U.S. space capabilities more resilient or less attractive as targets of attack, we define *defensive space control* more broadly than it is defined in Air Force and Joint space doctrine. In addition to active defenses, we include passive defenses and other approaches to defending, preserving, or restoring the capability under attack, such as dispersal to other space and terrestrial platforms, rapid replenishment, and rerouting of communications paths.

attempting to anticipate the other's future response and using that information to calculate the best current move. The game theory principles we apply in this methodology are those used in analyzing combinatorial games—sequential games in which each player has many moves from which to choose in each turn.

The methodology, illustrated in Figure S.1, employs sequential game theory in iterative analyses to

- determine the most likely ways "Red" (i.e., an adversary) would attempt to attack selected "Blue" (i.e., the United States) space capabilities
- assess which Blue DSC options would be most effective in defending against these attacks
- analyze whether Red would have viable options for countering Blue's defenses.

Figure S.1. Sequential Game-Theoretic Methodology

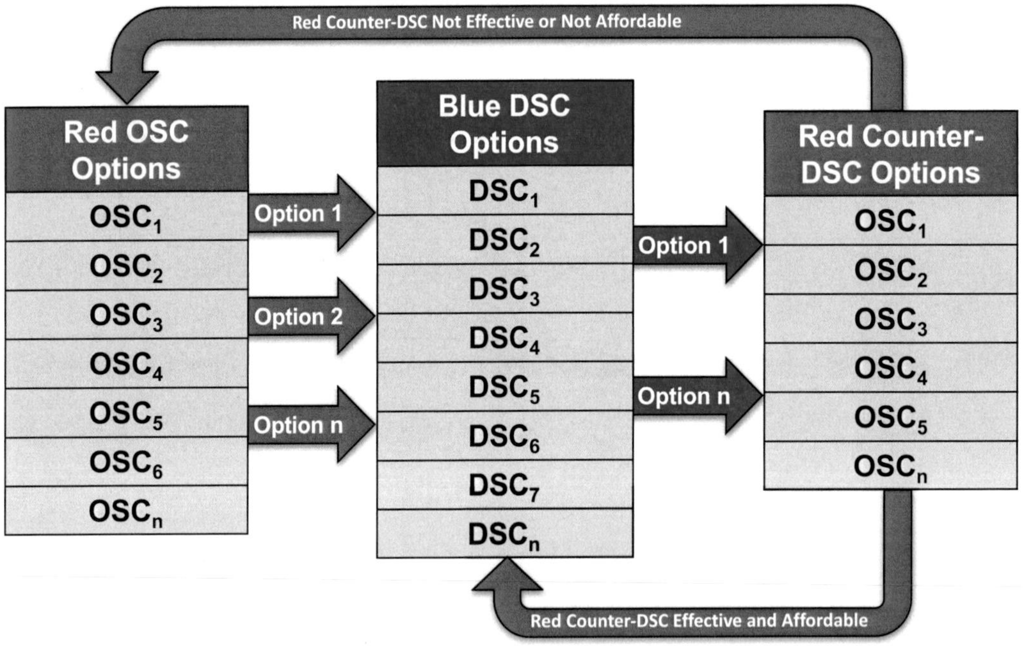

At each step, we evaluate an option according to its mission effectiveness, feasibility, likely policy cost, and potential to increase the risk of escalation. The iterative process illustrated in Figure S.1 continues until a dominant option is identified, a "saddle point" emerges, or the analysis results in an indeterminate draw. A dominant option exists when one side has an effective and affordable attack or defense with acceptable levels of escalation risk and political cost that the other side cannot effectively or affordably counter. If Red has a dominant option, Blue should expect Red to pursue its development, and Blue should intensify research and development (R&D) in search of a counter. If

Blue has a dominant option, then that is the capability Blue should develop, but Blue should expect that Red will continue to seek an effective counter. A saddle point exists when one side has an attack or defense that is highly effective, but its use would entail high escalation risk or political costs.[2] An indeterminate draw exists when both sides have only mediocre options available, none of them dominant. In either of the latter outcomes, Blue should focus its R&D on finding a route to a dominant option.

Each phase of the analysis is first driven by assessments of the potential mission effectiveness for each Red or Blue option. The effectiveness of Red OSC is measured by its ability to degrade Blue's space capabilities, while the effectiveness of Blue DSC is measured by its ability to restore Blue's space capability by countering the Red OSC. Data from external, high fidelity models, generally exercised by program offices, are expected to be the primary sources of mission effectiveness values. If model data are not available, inputs from subject-matter experts (SMEs) can also be used. Mission effectiveness is clearly a major driver, and less-effective options would be carried forward only if they have higher scores along the other criteria.

Next, feasibility assessments are entered. The methodology is designed to analyze potential Red-Blue interactions at future points in time, such as in 2020, 2025, or 2030. Only capabilities expected to be available to each side for when a particular analysis is being conducted should be assessed. Feasibility of OSC options for potential adversaries can be drawn from Intelligence Community assessments, which provide future dates and confidence levels. Program offices and research and development laboratories can assess the feasibility of potential Blue DSC approaches.

Then, the options deemed feasible by specified future dates are evaluated for escalation risk (which considers the views of Red and Blue only) and political cost (i.e., world opinion). Data for escalation risk and political cost are clearly more subjective than the other two criteria. We have developed quantitative baseline tables for escalation risk and political cost based on our expertise, but they can be modified to meet particular circumstances.

Scores for each of the four criteria (mission effectiveness, feasibility, escalation risk, and political cost) will illuminate deterrence prospects and potential escalation dynamics. For example, high mission effectiveness scores paired with high escalation risks suggest

---

[2] In formal game theory, the term *saddle point* describes the outcomes of simultaneous games played on a two-dimensional matrix in which "the larger of the row minima is equal to the smaller of the column maxima." Here, we use the term less formally to describe outcomes that simultaneously exhibit good mission effectiveness scores and poor escalation risk or political cost scores. See J. D. Williams, *The Compleat Strategyst: Being a Primer on the Theory of Games of Strategy*, New York: Dover Publications, Inc., 1986, p. 35.

pressure for conflict escalation. Options that are highly feasible and affordable with low political cost intensify these pressures. On the other hand, low mission effectiveness scores paired with high escalation risks suggest deterrence will probably endure until the conflict escalates for other reasons. If an option is feasible and cheap, Red might develop it, but will likely be deterred from using it unless the conflict escalates for other reasons. However, if feasibility or affordability is in question, Red will likely be dissuaded from attempting development.

Sensitivity to political cost will differ by Red actor and level of conflict. Some states are more sensitive to political cost than others. Pariah states have less to lose and will less likely be deterred by prospects of high political costs. Nations that are more concerned about reputation, such as the United States and other developed countries integrated in the global community, will more likely be deterred at low levels of conflict, but will be more willing to pay these costs if the conflict escalates. In any event, analyses employing the methodology can provide insights into what political costs leaders on each side should expect to pay in specific scenarios and may suggest ways to raise an opponent's political costs for certain kinds of attacks, thereby strengthening deterrence in selected areas.

Supporting Tools

The first step in determining what OSC and DSC systems each side might develop and employ is to identify the range of options available. While this is a complex problem, it can be approached systematically. One way to do this would be to develop what might be described as a Master Game Table (MGT), a cross-referenced spreadsheet mapping the full range of potential attacks on and defenses of space force enhancement systems. While using a fully populated MGT that maps the universe of possible attacks and defenses on space systems would be ideal, we recognize that developing such a comprehensive tool would be beyond the resource and knowledge constraints of some organizations.[3] Therefore, we have developed a separate tool that can be used to support analyses done within the game-theoretic framework without employing an MGT: the

[3] For the purposes of this study, we drafted a working-level MGT with non-program-specific information in it. Due to its limited utility at the level of generality at which it was developed, RAND decided not to publish this tool. RAND Project AIR FORCE encourages any organization that chooses to employ the game-theoretic methodology described in this report to develop an in-house MGT and populate it with detailed information on space force enhancement programs and prospective OSC and DSC options.

Defensive Space Analysis Tool (DSPAT).[4] Figure S.2 shows how the DSPAT works to provide information for decisionmakers.

Figure S.2. Methodology Inputs and Outputs

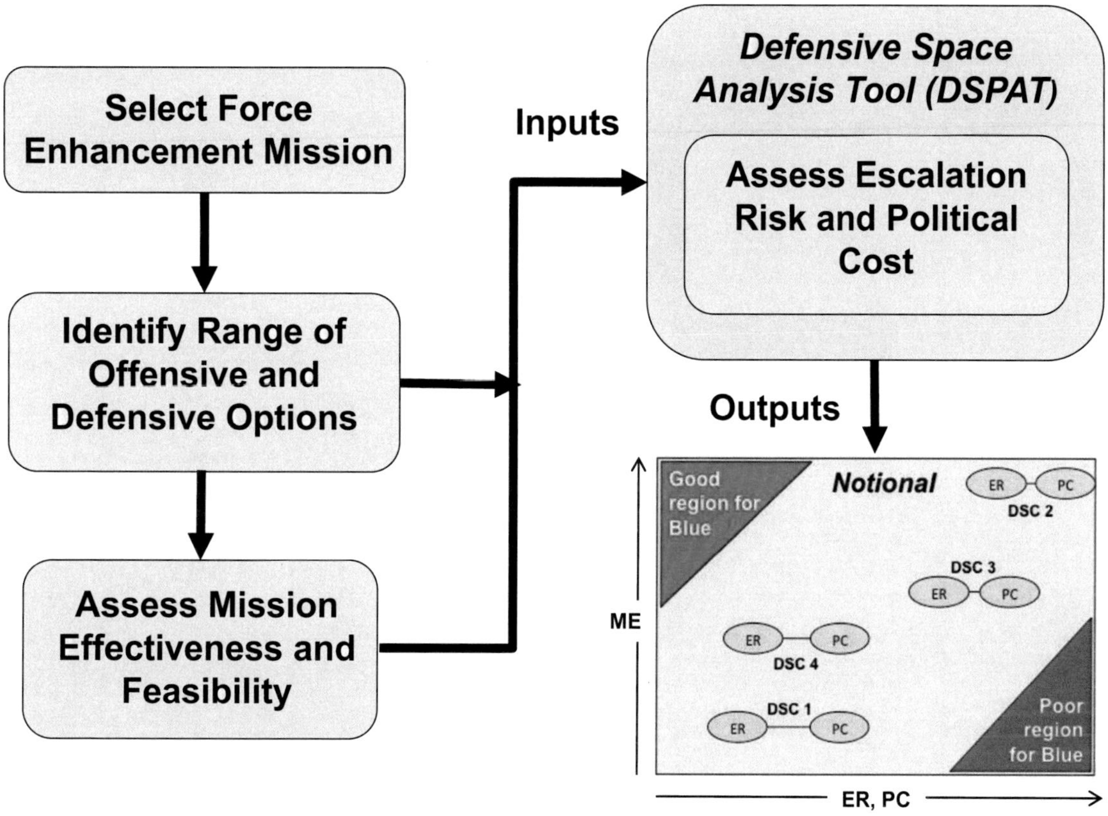

NOTE: ER = escalation risk; ME = mission effectiveness; PC = political cost.

The DSPAT is a decision support tool that solicits qualitative and quantitative inputs from SMEs, assigns relative values to the qualitative information it receives, and then submits those values to an integrated set of mathematical models that automate the scoring and ranking process in each of the four assessment categories. To begin a game-theoretic analysis, analysts first select a space force enhancement mission area to evaluate and identify some range of ways to attack that capability, or "offensive options," and a range of ways to defend against each of those attacks, or "defensive options." Then, analysts assess the mission effectiveness and feasibility of each of the offensive and

4 Ideally, the DSPAT would be used in conjunction with an MGT. They are mutually supportive.

defensive options under investigation and enter that information into the DSPAT. The DSPAT scores each option for escalation risk and political cost and ranks the outputs for decisionmaker review.

Figure S.3. Display of Notional DSC Analysis Outputs

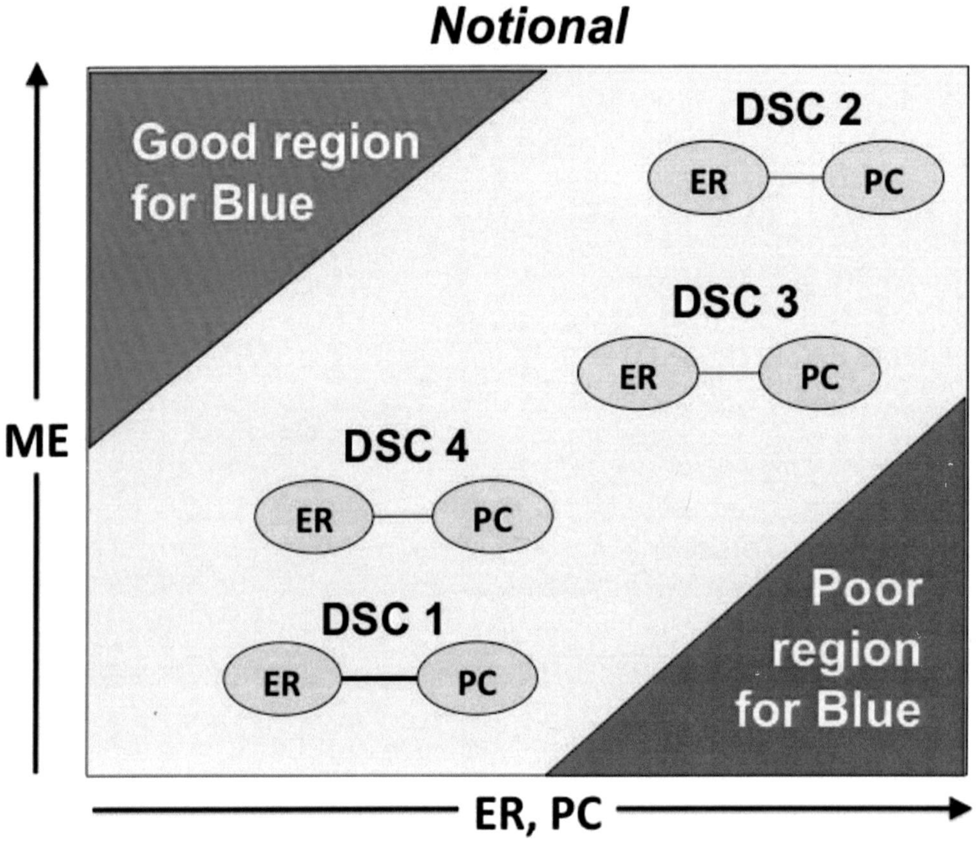

A notional outcome of the methodology is shown in the lower-right corner of Figure S.2, and this illustration is expanded in Figure S.3. In this example, four Blue DSC options were evaluated as possible ways to protect a Blue space capability from a specified Red OSC threat. Mission effectiveness (ME) (the ability to counter the Red OSC and restore Blue space capability) is plotted along the y-axis and escalation risk (ER) and political cost (PC) are plotted along the x-axis (only feasible options are displayed). Thus, placement in the upper-left corner would indicate high mission effectiveness and low escalation risk and political cost—a good option for Blue. Placement in the lower-right corner would indicate low mission effectiveness and high escalation risk and political cost—a poor option for Blue. In this example, DSC-2 represents a saddle point in which the option that scored best in mission effectiveness also scored highest in escalation risk and political cost. Had the options clustered in the middle of the graph, it would have indicated that the analysis ended in an indeterminate draw.

Types of Analysis

The game-theoretic methodology can be applied in four types of analysis to work through a variety of space deterrence and escalation management-related questions.

Dynamic Strategic Assessment

The type of analysis that most broadly applies the game-theoretic methodology can be described as a *Dynamic Strategic Assessment*. In this type, illustrated in Figure S.4, analysts choose a specific Blue space capability and evaluate the full range of ways to attack and defend it.

Figure S.4. Dynamic Strategic Assessment

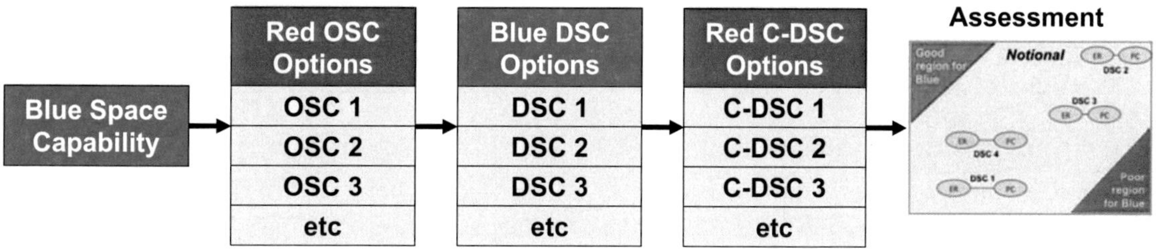

NOTE: C = counter.

Such comprehensive analyses of all ways to attack and defend Blue space capabilities in a specified scenario can reveal the most frequently predicted paths of conflict across the universe of possibilities, illuminating the levels of escalation and political cost that decisionmakers should expect during conflict in space.

Red Threat-Driven Assessment

Should intelligence reports indicate that a Red actor appears committed to developing a specific OSC capability, the game-theoretic methodology could be applied in an analysis specifically tailored to evaluate the implications of such a development and identify the most-desirable approach for defending against it. Figure S.5 illustrates the *Red Threat-Driven Assessment*.

Figure S.5. Red Threat-Driven Assessment

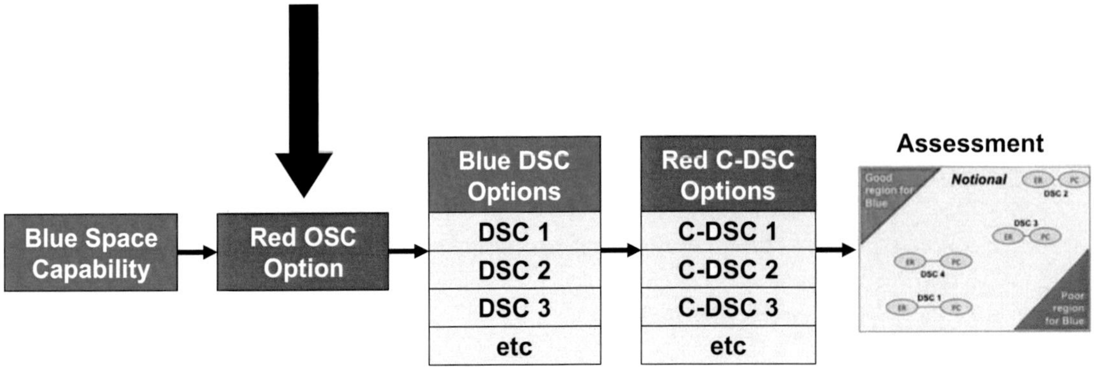

In this type of analysis, Red's preferred OSC option has become apparent, and one can deduce which Blue space capability it threatens. The analysis can dispense with the assessment of Red OSC options and proceed directly to searching for the best defense against the known threat. The analysis cycles between Blue DSC options and Red counter-DSC options until a dominant strategy, saddle point, or indeterminate draw is identified.

Blue Technology-Driven Assessment

From time to time, space program offices or system developers propose new systems for defending U.S. space capabilities or system enhancements for making them more robust against attack. Whenever such ideas are discussed, questions often arise as to how effective the proposed technology would be against which threats and the degree to which developing or employing it would be escalatory or politically costly.

Such questions can be addressed using the type of game-theoretic analysis shown in Figure S.6, the *Blue Technology-Driven Assessment*.

Figure S.6. Blue Technology-Driven Assessment

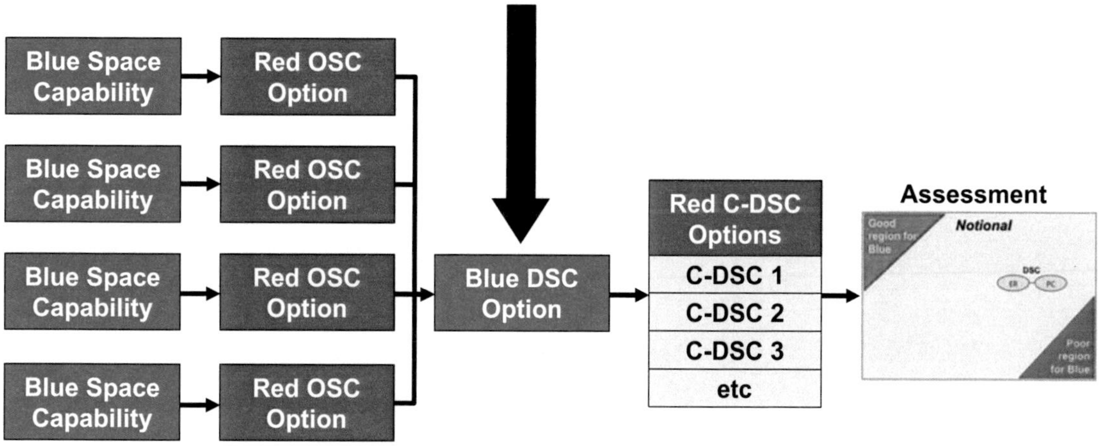

In this type, analysts assess the proposed Blue DSC option's mission effectiveness against a specified OSC attack, along with its feasibility, escalation risk, and political costs. Analysts also conduct an assessment of potential Red counter-DSC options. If they determine that the proposed DSC option would likely be effective against the specified OSC with acceptable escalation risks and political costs, they evaluate how effective it would be in defending against other potential threats. As Figure S.6 illustrates, a system that protects against four dangerous threats to U.S. space capabilities is more justifiable than one that defends against only a single threat, especially if the proposed system would be costly to develop.

Blue OSC Assessment

Although U.S. leaders are generally more interested in defending U.S. space capabilities than attacking those of potential adversaries, there are situations in which it might be advantageous to have certain Blue OSC capabilities available to posture in peacetime for deterrent purposes or employ in war for space control. As Figure S.7 illustrates, the game-theoretic analysis can be applied in a Blue OSC assessment to determine which Blue OSC capabilities would offer the highest mission effectiveness with manageable levels of escalation risk and acceptable political costs.

Figure S.7. Blue OSC Assessment

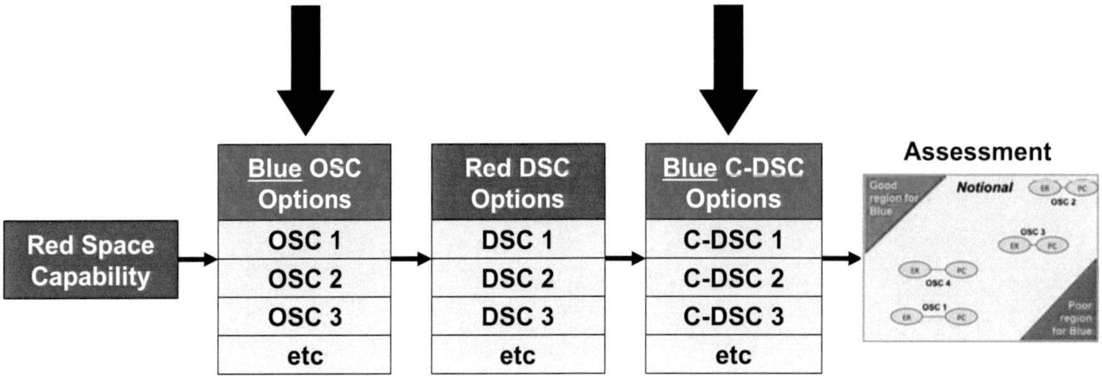

To conduct a Blue OSC assessment, analysts would determine which Red space capability U.S. leaders might want to attack. Then, they would develop a list of OSC options that could be used to attack that space capability and assess them for mission effectiveness, feasibility, escalation risk, and political cost. When the analysis identifies what appears to be the best OSC option for attacking the Red space capability, analysts could then conduct DSC assessments against it and counter-DSC assessments against the best DSC option that emerges, cycling through iterations of analysis, as described in the foregoing sections, until the best Blue OSC option is identified.

Applications and Recommendations

The game-theoretic methodology provides a broad framework for assessing the potential effects of alternative OSC and DSC systems. The breadth and versatility of this analytical framework suggest a wide range of potential applications. We discuss several here and offer recommendations for their implementation.

Support for Program Development Decisions

The methodology provides a means to assess the potential deterrence benefits and escalation risks of alternative OSC and DSC systems; therefore, program development decision support is its most obvious and direct application.

Recommendations

1. Make the DSPAT available to selected system program offices and other agencies involved in national security space system research and development, and encourage them to learn and apply the game-theoretic methodology.
2. Insist upon thorough, method-based analysis in briefings and papers advocating the development of any OSC or DSC system.

Support for Operational Training and Tactical Decisionmaking

The methodology can also make important contributions to ongoing national security space operations. The DSPAT can provide operational space commanders with valuable analytical support for interpreting the implications of intelligence assessments, anticipating opposing space force behavior during crises, and developing the most effective and appropriate courses of action when threats to U.S. space capabilities manifest. The MGT, if developed as a comprehensive survey of attacks and defenses, would also offer operators an important resource for operational training in space warfare.

Recommendations

1. Make the DSPAT available to the Joint Space Operations Center (JSpOC) and selected Air Force space operations centers.
2. Incorporate use of the game-theoretic methodology and other operational decisionmaking tools and procedures in the JSpOC Mission System and at Air Force space operations centers.
3. Develop a comprehensive MGT and incorporate information from it in training programs at Air Force space operations units and centers, and train crew commanders on the tactical employment of method-based analysis.

Support for Space Play in War Games

The game-theoretic methodology can bring a higher level of coherence and objectivity to space play in war games. Game developers can use it to assess what Red OSC attacks would best support the scenario being designed and anticipate what would be good and bad Blue team responses to those attacks. Players on both sides could use the DSPAT to compare the escalation risks and political costs of space control options before deciding on appropriate courses of action. Adjudicators could assess Red and Blue inputs against each other in DSPAT and estimate space conflict outcomes, along with associated levels of escalation and political cost, with higher levels of analytical rigor than is possible with purely subjective assessments.

Recommendations

1. Make the game-theoretic methodology and any supporting tools available to Air Force, Joint, and Office of the Secretary of Defense (OSD) war game developers.
2. Encourage war game developers and controllers to make the MGT and DSPAT available to players.
3. Use the game-theoretic methodology and any supporting tools to support space event adjudication in major war games.

Support for Strategic Planning and Policymaking

One of the most important features of the game-theoretic methodology is its ability to do Dynamic Strategic Assessments of what OSC and DSC actions each belligerent would likely take at each level of confrontation and conflict in specified scenarios. These assessments can inform force structure decisions, doctrine and strategy development, and policy development for maximizing strategic stability and protecting national interests.

Recommendations

1. Make the game-theoretic methodology and supporting tools available to Air Force and Joint planners, space policymakers in OSD, and planning and policymaking offices in other national security space-related departments and agencies.
2. Conduct systematic, comprehensive Dynamic Strategic Assessments to determine likely paths of conflict in space in all defense planning scenarios and other scenarios of interest as they arise.

Acknowledgments

We would like to thank our action officer, Lt Col Bryan "Stu" Eberhardt (Space and Missile Systems Center/SYAF), and his staff for their support and guidance.

Abbreviations

ADF	Australian Defence Force
AFB	Air Force base
AFSPC	Air Force Space Command
ASAT	anti-satellite
CHIRP	Commercially Hosted Infrared Payload
DARPA	Defense Advanced Research Projects Agency
DoD	Department of Defense
DSC	defensive space control
DSPAT	Defensive Space Analysis Tool
EMI	electromagnetic interference
ER	escalation risk
F6	Future, Fast, Flexible, Fractionated, Free-Flying Spacecraft United by Information Exchange
FOBS	Fractional Orbital Bombardment System
GEO	Geosynchronous Earth Orbit
GPS	Global Positioning System
HEO	Highly Elliptical Orbit
IOC	initial operational capability
ISR	intelligence, surveillance, and reconnaissance
JMS	JSpOC Mission System
JSpOC	Joint Space Operations Center
LEO	Low Earth Orbit
ME	mission effectiveness
MEO	Medium Earth Orbit
MGT	Master Game Table
NGA	National Geospatial-Intelligence Agency
NRO	National Reconnaissance Office
NSS	national security space
ORS	Operationally Responsive Space
OSC	offensive space control
OSD	Office of the Secretary of Defense
PAF	RAND Project AIR FORCE
PC	political cost
PNT	positioning, navigation, and timing

R&D	research and development
SAR	synthetic aperture radar
SME	subject-matter expert
SSA	space situational awareness
UHF	ultrahigh frequency

1. Introduction

Contested Space

U.S. space capabilities are at risk. Potential adversaries are motivated to develop offensive space control (OSC) systems. They are well aware that U.S. transformational warfighting capabilities are becoming increasingly reliant on support from space. This is especially true for expeditionary operations, which, for example, rely heavily on satellite communications for intra-theater communications and obtaining products and services from organizations that are not forward deployed (i.e., "reachback").[1] Given this knowledge, potential adversaries reason that possession of OSC capabilities may deter the United States from conducting military operations in their sphere of influence, and would enable them to degrade U.S. space capabilities if the United States is not deterred. For example, a January 2013 article in a Chinese newspaper owned by the Chinese Communist Party states that

> … it is necessary for China to have the ability to strike US satellites. This deterrent can provide strategic protection to Chinese satellites and the whole country's national security.[2]

Moreover, potential adversaries may feel justified in developing OSC capabilities, believing that the United States has already done so and is continuing down that path. First, they can point to past U.S. space policies. Declassified records and White House fact sheets reveal that U.S. national space policies from the Reagan administration onward have stated that the United States will develop and, if necessary, employ capabilities to impose space control and conduct force application from space.[3] The fact sheet for the 2006 U.S. National Space Policy, for instance, states that the United States will "deny, if necessary, adversaries the use of space capabilities hostile to U.S. national

---

[1] In contrast, potential adversaries are not as dependent on their space assets if U.S. expeditionary operations are near their homelands.

[2] "Satellite Test Sparks Overblown Worries," *Global Times,* January 6, 2013.

[3] White House, "Presidential Directive on National Space Policy," unclassified summary, Washington, D.C., February 11, 1988; National Space Policy Directive-1, *National Space Policy Directives and Executive Charter,* Washington, D.C.: White House, November 2, 1989; National Science and Technology Council, "Fact Sheet: National Space Policy," White House, Office of Science and Technology Policy, Washington, D.C., September 19, 1996; White House, "U.S. National Space Policy," Washington, D.C., August 31, 2006; and White House, *National Space Policy of the United States of America*, Washington, D.C., June 28, 2010.

interests."[4] A Chinese military academician has opined that this policy "pursues hegemony in space [by the United States] and poses a significant security risk to China that cannot be left unaddressed."[5] Potential adversaries can then conclude that the United States is carrying out this policy, by pointing to past anti-satellite (ASAT) programs,[6] recent controversial activities that have "offensive" characteristics,[7] and one current program.[8] Not only can potential adversaries use such activities to justify their own, they may feel compelled do so to appear strong to their domestic constituents, their immediate neighbors, and the international community.

Given the above arguments, it should not be surprising that China demonstrated a direct-ascent ASAT capability against one of its satellites on January 11, 2007, a fact that Beijing confirmed 12 days later following international outcry about the large amount of space debris the test caused.[9] While this has been the most visible and provocative act to date, it represents only a portion of China's space control program efforts. According to the *2015 Annual Report to Congress of the U.S.-China Economic and Security Review Commission*:

> China is pursuing a broad and robust array of counterspace capabilities, which includes direct-ascent antisatellite missiles, co-orbital antisatellite systems, computer network operations, ground-based satellite jammers, and directed energy weapons. China's nuclear arsenal also provides an inherent antisatellite capability.[10]

---

[4] White House, 2006.

[5] Bao Shixiu, "Deterrence Revisited: Outer Space," *China Security*, No. 5, Winter 2007, pp. 2–11.

[6] Examples of such programs include Program 505 nuclear-armed Nike Zeus missile, Program 437 nuclear-armed Thor missile, and the F-15A-launched interceptor with a miniature homing vehicle for kinetic kill.

[7] Examples include Orbital Express, a system that demonstrated robotic, autonomous on-orbit satellite servicing; XSS-10 and XSS-11, two experimental satellite systems that demonstrated autonomous rendezvous and proximity maneuvers; the Phoenix program, whose goal is to develop and demonstrate technologies to cooperatively harvest and reuse valuable components from retired, non-working satellites; the shoot-down of USA 193, a non-functioning U.S. satellite that contained toxic fuel, by a modified Standard Missile 3; and X-37B Orbital Test Vehicle, an ongoing program used to test new space technologies, but which a newspaper funded by the Chinese government has called a "space fighter." See Zhengyan Fang, "America's Space Fighter: New Threat to Humanity," *Ta Kung Pao*, May 5, 2010.

[8] That program, Counter Communication System, is a ground-based radio frequency jammer and the only OSC capability acknowledged by the United States.

[9] "China Confirms Satellite Downed," *BBC News*, January 23, 2007.

[10] William A. Reinsch and Dennis C. Shea, *2015 Annual Report to Congress of the U.S.-China Economic and Security Review Commission*, 114th Congress, First Session, Chapter 2, Section 2, Washington, D.C.: U.S. Government Printing Office, p. 292. Interestingly, another nation is using China's 2007 ASAT demonstration to justify its own efforts. A senior military officer in India stated that "our satellites are vulnerable to ASAT weapon systems because our neighbourhood possesses one," while expressing the

Defense and Deterrence

Left unchallenged, such developments increase the chances that attacks on U.S. space systems will become more common. While a nonreversible kinetic ASAT capability is clearly a more serious threat than OSC capabilities that have reversible effects, U.S. space systems have, in fact, already been the target of these latter systems. For example, adversaries have used ground-based jammers to interfere with the Global Positioning System (GPS) and satellite communications during recent conflicts in Southwest Asia.[11] The frequency and severity of attacks on U.S. space capabilities are likely to increase in future confrontations as opponents develop more capable OSC technologies. As stability in space erodes, crisis stability in the terrestrial domain could suffer as potential adversaries become more emboldened to challenge U.S. interests, believing they can either deter U.S. intervention by threatening space assets or defeat U.S. forces after interdicting their support from space.[12] Therefore, it behooves the United States to develop appropriate responses to emerging threats to its space capabilities.

The 2010 *National Space Policy* and 2012 *Department of Defense Space Policy* call for a number of activities to defend U.S. space capabilities and deter potential adversaries from attacking them. OSC and defensive space control (DSC) systems are among the approaches under consideration. Space control systems can contribute to deterrence by demonstrating that attacks on U.S. space systems can be withstood or defeated and that the United States can also hold enemy space systems at risk. But deciding which space control systems to develop is a complex problem. Some are more effective than others, and some may be escalatory, increasing risks of even greater losses, or provocative, generating political costs for the United States.

While a general discussion of deterrence would involve all of the world's space-faring nations, the United States should focus its attention on those potential adversaries that have the resources to develop space control capabilities and whose past actions indicate they intend to do so. At this time, the number of such nations is low, and the United States should have a goal of keeping that list short. Most of the other space-faring

---

need for India to develop ASAT technology. See Bhrarath Gopalaswamy and Harsh V. Pant, "Does India Need Anti-satellite Capability?" *Rediff News*, February 9, 2010.

[11] Donna Miles, "Iraq Jamming Incident Underscores Lessons About Space,"*American Forces Press Service*, September 15, 2004; Lt Gen Dan P. Leaf, "Providing Combat Effects to the Battlefield," *High Frontier*, Vol. 1, No. 4, 2005, pp. 4–5; LTG Larry J. Dodgen, "Leveraging Space to Support the Changing Paradigm," *High Frontier*, Vol. 1, No. 4, 2005, pp. 6–11.

[12] For an insightful discussion on the relationship between space deterrence and general deterrence, see James P. Finch and Shawn Steene, "Finding Space in Deterrence: Toward a General Framework for 'Space Deterrence,'" *Strategic Studies Quarterly*, Winter 2011.

nations, such as those in Europe, are very concerned about space control developments and are actively working to establish an international code of conduct for outer space activities.[13] Codes of conduct and other international agreements can enhance stability in the space domain by raising political costs to states that violate them. But the specter of political censure alone may not be enough to deter adversaries from attacking U.S. space capabilities in a time of crisis or conflict. The United States needs to take more direct measures to defend its space systems and deter attacks on them.

The Air Force plays a key role in defending U.S. space assets and can play a key role in deterrence by developing and operating space systems that are more resilient and can demonstrate U.S. capability and intent to defend these systems to potential adversaries. In particular, DSC is Air Force Space Command's (AFSPC's) fifth space priority of 17 space priorities.[14] DSC demonstrations may be effective in convincing a potential adversary that its attacks could be easily nullified.[15] As a result, the adversary may be deterred from employing its OSC systems during crisis or conflict, or, better yet, be dissuaded from pursuing the development of such systems in the first place.

Objective

The objective of the research reported in this report is to help the Air Force assess the deterrent value of alternative DSC options. Specifically, the research sought to develop a methodology to identify those DSC options that would likely contribute to deterrence without increasing political costs and risks of escalation. One benefit of a formal methodology is that it can provide a systematic framework with standard criteria for analysts. It also can provide more persuasive support for program advocacy by providing analysis to justify assertions regarding deterrence prospects and escalation risks of proposed systems.

---

[13] Council of the European Union, "Code of Conduct for Outer Space Activities," revised draft, Brussels, October 11, 2010.

[14] "Air Force Space Command Defines Its Priorities," *Air Force Magazine,* July 17, 2012; and AFSPC, "Space and Cyberspace Priorities," web page, undated.

[15] Note that deterrence requires that selective information about U.S. DSC capabilities be releasable, either publicly or via diplomatic channels, to potential adversaries. If DSC capabilities are not obviously achievable, they need to be demonstrated in an easily verifiable manner. Finally, the United States must signal its intent to deploy and employ DSC capabilities (this is likely less important as the international community may already believe this to be true).

Approach

The research in this report involved a four-step process. First, we reviewed U.S. national-level and U.S. Department of Defense (DoD)-level space policy and strategy documents for DSC guidance, direction, and stance on deterrence. Next, we identified a range of OSC options that could threaten U.S. space systems. We reviewed the literature to create a list of possible near-term and far-term OSC systems. We then identified a range of U.S. DSC options, both near term and far term. The OSC and DSC options were not modeled after specific systems, but the concepts were characterized sufficiently for subsequent analysis. Finally, we developed a methodology, based on sequential game theory, to assess the deterrent value, escalation dynamics, and potential political costs of the various DSC options. This methodology is very versatile. Not only does it allow analysts to assess which DSC approaches are the most promising against specific OSC threats, it also enables them to assess which OSC capabilities potential adversaries are likely to develop and employ, evaluate the benefits and risks of any OSC capabilities that U.S. leaders might want to develop, and anticipate what paths of conflict and patterns of escalation might emerge in space given alternative OSC and DSC force structures on each side.

Report Structure

Chapter Two provides a brief review of U.S. national and DoD space policies and strategies, with a focus on DSC and deterrence. Chapters Three through Five then explain the game-theoretic methodology, the types of analysis to which it can be applied, and how to interpret its outcomes. Chapter Six provides summary observations and some recommended applications of the methodology.

2. Policy Guidance

This chapter provides a brief review of U.S. national and DoD space policy and strategies, with a focus on deterrence, resilience, and DSC to set the context for the research.

National Space Policy

The 2010 *National Space Policy of the United States of America* provides guidance on the need for mission assurance, resilience, deterrence, and self-defense. Specifically, the United States will adhere to five principles, with the fifth being:

> The United States will employ a variety of measures to help [ensure] the use of space for all responsible parties, and, consistent with the inherent right of self-defense, deter others from interference and attack, defend our space systems and contribute to the defense of allied space systems, and, if deterrence fails, defeat efforts to attack them.[1]

Consistent with the five principles, the United States will pursue six goals, with the fourth being:

> Increase assurance and resilience of mission-essential functions enabled by commercial, civil, scientific, and national security spacecraft and supporting infrastructure against disruption, degradation, and destruction, whether from environmental, mechanical, electronic, or hostile causes.[2]

National Security Space Strategy

The 2011 *National Security Space Strategy* expands on the 2010 *National Space Policy* by charting a path for the next decade for the national security space (NSS) community to respond to the current and projected space strategic environment. Specifically, the United States will pursue a number of strategic approaches to meet its NSS objectives, including a multilayered approach to dissuade and deter the development, testing, and employment of space control systems.[3] The elements include the following:

---

[1] White House, 2010, p. 3.

[2] White House, 2010, p. 4.

[3] Department of Defense and Office of the Director of National Intelligence, *National Security Space Strategy: Unclassified Summary*, Washington, D.C., January 2011, p. 10.

- support diplomatic efforts to promote norms of responsible behavior in space
- pursue international partnerships that encourage potential adversary restraint
- improve our ability to attribute attacks
- strengthen the resilience of our architectures to deny the benefits of an attack
- retain the right to respond, should deterrence fail.

Resilience and DSC

As *resilience* is a relatively new term to the space domain, the Office of the Deputy Assistant Secretary of Defense for Space Policy has provided a definition of the term:

> Resilience is the ability of an architecture to support the functions necessary for mission success in spite of hostile action or adverse conditions.[4] An architecture is "more resilient" if it can provide these functions with higher probability, shorter periods of reduced capability, and across a wider range of scenarios, conditions and threats. Resilience may leverage cross-domain or alternative government, commercial, or international capabilities.[5]

As Table 2.1 illustrates, DSC is one of many "system" options[6] to increase the resilience of U.S. space architectures. Decisionmakers must weigh all options and select the optimal combination, balancing effectiveness and affordability.

---

[4] Adverse conditions include acquisition delays, early on-orbit failure, space environment effects, collision with orbital debris, as well as enemy action.

[5] U.S. Department of Defense, "Resilience of Space Capabilities," fact sheet, Washington, D.C., 2011; Department of Defense Directive 3100.10, *Space Policy*, Washington D.C., October 18, 2012, incorporating Change 1, effective November 1, 2016.

[6] As opposed to non-material options, such as diplomacy.

Table 2.1. Options to Increase Resilience of U.S. Space Systems

Category	Examples
Proliferation	• International cooperation (i.e., leverage foreign space assets, which can be relabeled Operationally Responsive Space [ORS] Tier 1) • U.S. commercial sources (e.g., imagery from DigitalGlobe and GeoEye) • Foreign commercial sources (e.g., National Geospatial-Intelligence Agency [NGA] contracts with foreign synthetic aperture radar [SAR] providers) • Hosted payloads (e.g., Commercially Hosted Infrared Payload [CHIRP] on SES-2, Australian Defence Force [ADF] ultrahigh frequency [UHF] payload on Intelsat-22)
Rapid reconstitution	• AFSPC's Rapidly Deployable Space (ORS Tier 2)
New architectures	• Force mix: both large and small satellites • Fractionated concepts: Space-Based Group, Defense Advanced Research Projects Agency (DARPA), System F6
DSC, including space situational awareness (SSA) as an enabler	• Air Force/National Reconnaissance Office (NRO) Space Protection Program • Self-awareness SSA • Cost-effective protection (hardening, link encryption, maneuverability, ground facility protection, electromagnetic interference [EMI] characterization and geolocation, increased signal strength, and others) • Active measures
Cross-domain solutions	• Capabilities from the air, land, sea, and cyber domains

SOURCES: U.S. Department of Defense, Air Force, and contractor fact sheets.

We note that AFSPC considers DSC important by making it the command's fifth space priority in this list:[7]

- Nuclear, survivable communications
- Launch detection/missile tracking
- Position, navigation, and timing
- SSA and battlespace awareness
- DSC
- Ensure space access/spacelift
- Space command and control
- Satellite operations
- Protected, tactical communications
- OSC
- Unprotected communications
- Space-to-surface ISR

[7] "Air Force Space Command Defines Its Priorities," *Air Force Magazine*, July 17, 2012; AFSPC, 2012.

- Terrestrial Environmental Monitoring
- Nuclear Detonation Detection
- Responsive Spacelift.

DoD Space Policy

The 2012 DoD *Space Policy* reiterates the statements in the 2011 *National Space Policy*, but adds two interesting details: that space interference can be "escalatory" and that the United States will respond at the time and place of its choosing:

> The sustainability and stability of the space environment, as well as free access to and use of space, are vital to U.S. national interests. Purposeful interference with U.S. space systems, including their supporting infrastructure, will be considered an infringement of U.S. rights. Such interference, or interference with other space systems upon which the United States relies, is irresponsible in peacetime and may be escalatory during a crisis. The United States will retain the capabilities to respond at the time and place of our choosing.[8]

For this reason, it is important to analyze DSC options not only for their mission effectiveness, but also for the political costs and escalation risks involved in using them. This is a key feature of the RAND model discussed in the following chapter.

[8] Department of Defense Directive 3100.10, 2012, p. 1.

3. Explaining the Methodology

Framing a Complex Problem

Deterring attacks on U.S. space capabilities is a complex challenge. The United States uses multiple space systems to enhance the warfighting effectiveness of its terrestrial forces. These systems operate in several orbital regimes and employ a variety of engineering approaches to accomplish their missions; therefore, each of them presents particular strengths and vulnerabilities to potential enemies contemplating ways to interrupt the force enhancement services they provide. Given this diversity in U.S. systems and their attributes, potential opponents have a variety of options available to them for attacking U.S. space capabilities. Each of these OSC options, in turn, exhibits its own strengths and vulnerabilities, suggesting multiple possibilities for developing systems to defend U.S. space capabilities from attack.

U.S. leaders need to identify and develop DSC systems that make their space capabilities as resilient as possible within budgetary constraints. A resilient space architecture, one that can withstand or defeat enemy attack and continue enhancing U.S. warfighting capabilities, helps deter potential adversaries from employing certain OSC systems, given the potential costs of U.S. retribution and the escalation risks and political costs associated with those systems. If those OSC options are also difficult or costly to develop, some opponents might even be dissuaded from attempting to do so. In sum, carefully chosen investments in DSC capabilities could have important payoffs in crisis stability, escalation management, and the combat effectiveness of U.S. terrestrial forces.

But how can U.S. leaders determine which DSC capabilities to develop, given the uncertainties surrounding which OSC systems potential opponents might pursue and the wide variance in mission effectiveness, feasibility, escalation risk, and political cost across the array of DSC options?

This chapter presents a logical framework and analytical process for working through this complex problem in a structured manner. Based on the "look forward and reason back" methodology employed in sequential game theory, analysts can identify which combinations of OSC and DSC options are most likely to be developed and employed and which, if any, will likely dominate in various encounters. The analysis begins with mapping a range of potential attacks on one or more space force enhancement capabilities, identifying possible defenses against those attacks, and considering conceivable counters to those defenses. Once accomplished, analysts are led through assessments of mission effectiveness, feasibility, escalation risk, and political cost for each OSC and DSC option. Then, iterative cycles of analysis illuminate deterrence and dissuasion prospects and potential escalation dynamics in alternative OSC and DSC matchups. RAND Project AIR FORCE (PAF) has developed a tool, which we describe in this chapter, to guide analysts through the various assessments and score alternative OSC and DSC approaches: the Defensive Space Analysis Tool (DSPAT). The appendix provides a fuller explanation of how to employ this tool.

Sequential Game Theory and the Analysis of Space Control Options

Sequential game theory offers a logical framework for anticipating what attacks and defenses adversaries are likely to develop for employment at various levels of conflict. It is a structured methodology applied to situations in which opponents make moves at different times, or in turn, and those moves and are interdependent—i.e., how much each side benefits or loses from its move depends on how the other side acts or reacts. As each opponent can anticipate how the other side could move (and knows the other side can too), identifying the best possible strategy in such a game requires each actor to look ahead to see how its actions would influence the future actions of others and adjust its moves accordingly. In other words, each side must *look forward and reason back.*[1]

A Tool for Solving Simple Problems—The Game Tree

A common way of working through a sequential game is by mapping out all possible moves in a game tree and scoring the outcomes. Figure 3.1 offers an example of how a game tree can be used to solve a simple strategy problem.

[1] An excellent tutorial on the look forward and reason back methodology is provided in Avinash K. Dixit and Barry J. Nalebuff, *Thinking Strategically: The Competitive Edge in Business, Politics, and Everyday Life*, New York: W. W. Norton & Co., 1991.

Figure 3.1. Attack or Defend Game Tree

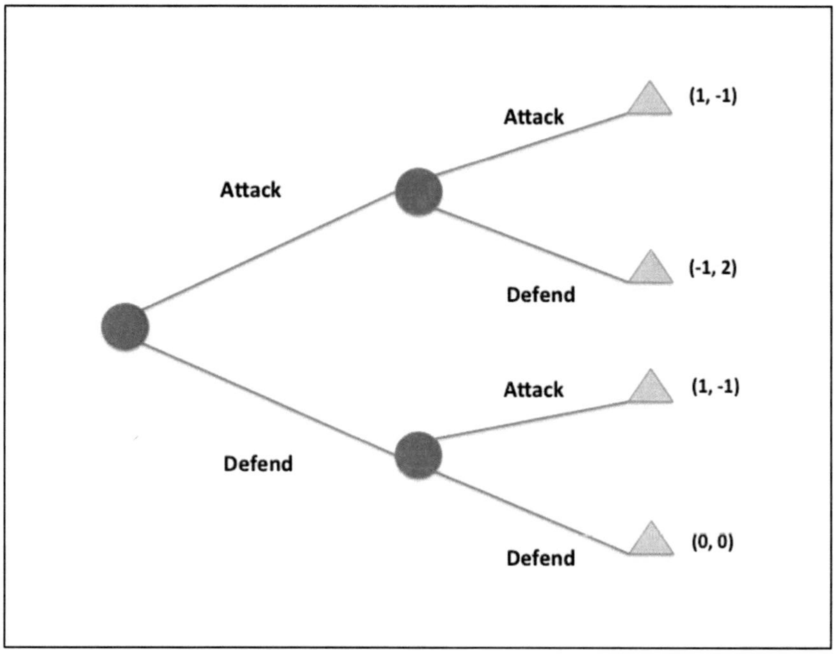

In this example, the king of a notional state, Red, covets the territory of his neighbor, Blue, and has mobilized his army on the Blue border, creating a crisis. The king of Red knows that his army is superior to Blue's and is confident in its ability to prevail if Blue will meet him on the open field of battle. However, the king of Red is unsure whether he can bring Blue to battle or, if he cannot, whether he should attack Blue in its defensive positions. The queen of Blue, in turn, knows a superior force is threatening her kingdom. The questions each side must consider are: Should either of them attack; and will the crisis remain stable or devolve to war?

The game tree in Figure 3.1 solves this problem by identifying each actor's decision nodes and resultant payoffs. If Red can lure Blue onto the open field, it can win a decisive victory, a score of 1, at Blue's expense, giving Blue a score of –1. However, Red must look ahead and anticipate that the queen of Blue knows her army would be defeated on the field, so she will keep it in its strong defensive positions where it would score 2 against a Red attack (earning her an extra point in winning a righteous cause) and deal Red a defeat, a score of –1. With this knowledge, the king of Red reasons back and is deterred from attacking. Similarly, the queen of Blue knows if she attacks the Red army on her border, she will likely be defeated, receiving a score of –1 and giving Red a score of 1. Thus both sides are deterred and the crisis remains stable—at least until Red devises a strategy or technology to undermine Blue's defense or Blue develops military forces more capable of fighting in the field.

Approaches and Tools for More Complex Problems

While the foregoing illustration is informative, space control strategy problems are too complicated to work using simple game trees. Game trees can map and score a few options at each player's decision nodes, but in the world of OSC and DSC operations, multiple space force enhancement capabilities must be defended, and there are many ways to attack each of those capabilities. There are, in turn, many alternative approaches for defending against each of those attacks. Moreover, some DSC options can be countered, nullifying their value, and the time it takes the effects of each OSC, DSC, and counter-DSC option to manifest must also be taken into account. Finally, as mentioned earlier, each OSC and DSC option must be scored in four dimensions—mission effectiveness, feasibility, escalation risk, and political cost—rendering simple two-dimensional trees inadequate for mapping the complex choices decisionmakers are forced to navigate.

Although basic sequential game theory is overtaxed by the complexity of such strategy problems, an advanced branch of sequential game theory devoted to combinatorial games is well suited to dealing with them. Combinatorial game theory is a branch of applied mathematics and computer science that studies sequential games in which high numbers of moves are possible in each turn.[2] In pure combinatorial games, such as the board games chess and Go, well-defined rules establish all possible moves, the payoffs (or costs) of each encounter, and the conditions of victory. Such perfect information is never present in the global strategic environment, but the fundamental logic of combinatorial game theory can be applied to real-world strategy problems to the extent that one can define the bounds of the problem under investigation, identify the universe of possible moves and countermoves within those boundaries, and assign relative payoffs to conditional outcomes of those moves in each of the dimensions being analyzed.[3]

---

[2] John H. Conway developed combinatorial game theory in the mid-1970s and publicized it in his now classic text, *On Numbers and Games*, London: Academic Press, 1976. Other important works in this field include Elwyn Berlekamp, John H. Conway, and Richard Guy, *Winning Ways*, Vols. 1–2, London: Academic Press, 1982; and Donald E. Knuth, *Surreal Numbers*, Reading, Mass.: Addison-Wesley, 1974.

[3] See Eric D. Demaine and Robert A. Hearn, "Playing Games with Algorithms: Algorithmic Combinatorial Game Theory," in Michael H. Albert and Richard J. Nowakowski, eds., *Games of No Chance 3*, Vol. 56, Mathematical Sciences Research Institute Publications, Cambridge, UK: Cambridge University Press, 2009, pp. 3–56.

Figure 3.2 illustrates the analytical framework for applying combinatorial game theory to analyze the potential dynamics of OSC and DSC confrontations.

Figure 3.2. Game Theory Framework for Space Control Analysis

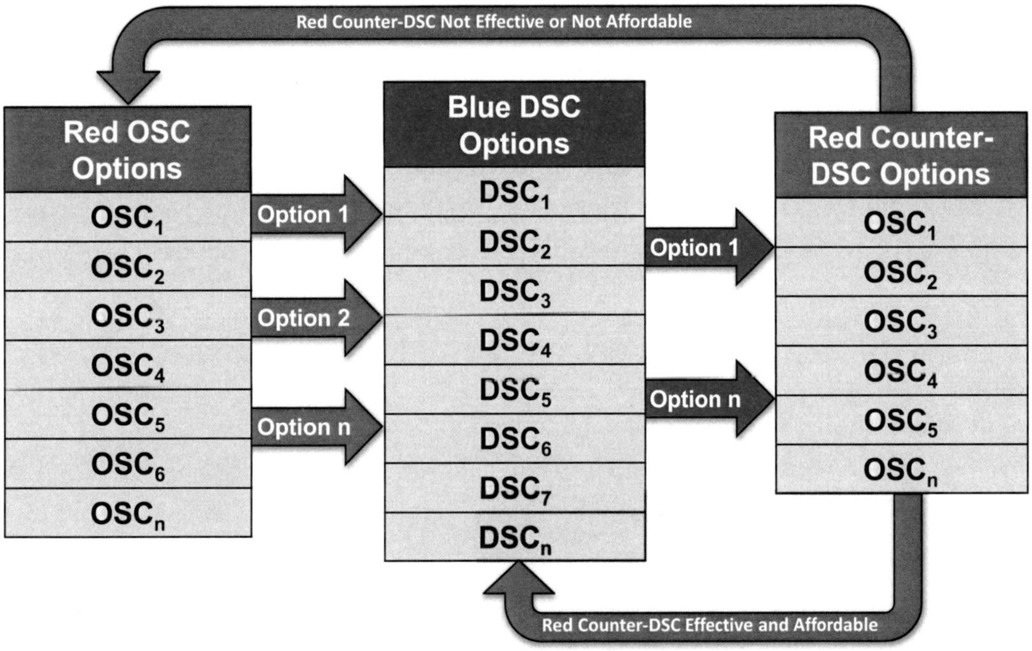

In this example, a notional actor, Red, is interested in interdicting a specific (although unspecified here) space capability that his opponent, Blue, is using to enhance the warfighting effectiveness of his military forces. To assess which space control options each actor might employ should a Red-Blue conflict occur in some future year, one would begin by identifying and listing all the ways Red could interdict the Blue capability, i.e., all Red OSC options, and ranking them on the basis of mission effectiveness, feasibility, escalation risk, and political cost.[4] The most-effective OSC option believed feasible by the designated year is selected as Red's preferred option, Option 1. Next, the analyst would identify and list all the ways Blue could defend its force enhancement capability against Red Option 1 and rank them in each of the aforementioned dimensions. From among the feasible Blue DSC options, the one determined to be most capable of reducing the effects of Red Option 1 is identified as Blue's preferred option, Blue Option 1. Then, the analysis would consider whether Red

[4] We shall explain the methods and criteria used in evaluating and ranking these dimensions later in this chapter.

has any effective and affordable means of countering Blue Option 1, thereby restoring the lost effectiveness of its original attack. If such means are available, Blue must anticipate that Red would develop and employ that counter, so Blue must discard Option 1 and consider its next most-desirable defense, Option 2. On the other hand, if Red is not expected to have a counter-DSC option available in the designated time frame that is both effective and affordable, it must anticipate that Blue would develop and employ such a defense. Therefore, Red must discard OSC Option 1 and consider its next most-desirable attack, OSC Option 2. Blue must then identify and assess its DSC Options against Red Option 2.

In each of these assessments, analysts must consider the time it would take each OSC, DSC, and counter-DSC option to take effect and adjust the rankings accordingly. If it would take hours, days, or longer to analyze and counter an OSC attack that could manifest in seconds, the excessive period of mission loss incurred would indicate that the DSC under consideration is less than desirable, even if it is 100-percent effective once it manifests. This is particularly true if a counter-DSC could be quickly employed or the OSC attack could be quickly adjusted and renewed, setting back Blue mission effectiveness for another long period. In such a case, analysts should seek a timelier defense, even if it is somewhat less effective than the original one when fully manifested.

In any case, the analysis proceeds in cyclic fashion, working down the ordered lists of OSC, DSC, and counter-DSC options until one of several possible outcomes emerges. The first possibility is that a dominant strategy will appear for one side or the other. That is, either (1) an effective Red OSC option will emerge with acceptable levels of escalation risk and political cost for which Blue has no feasible, effective counter, thus putting the threatened Blue force enhancement capability at critical risk; or (2) one or more Blue DSC options will emerge that, singularly or in combination, with acceptable levels of escalation risk and political costs, defeat all Red OSC options, thus making the threatened Blue space force enhancement capability unassailable.[5] A second possibility is that the analysis will reach a "saddle point" for one side or the other in which the best feasible OSC or DSC option for mission effectiveness is also the most dangerous for escalation risks or political costs. Finally, a third possibility is that the analysis will descend into an indeterminate draw, one in which each side has some number of space control options available, all offering moderate levels of mission effectiveness with moderate escalation risks and political costs.

---

[5] As we shall explain later, what constitutes "acceptable levels" of escalation risk and political cost varies across levels of conflict and the risk tolerances and political sensitivities of the actors involved.

All of these outcomes are informative. Each has implications for defense and deterrence, strategic stability, probable paths of conflict and levels of escalation, and therefore investment and force posture decisions. We will explain how to interpret various outcomes in Chapter Five. The discussion continues below with a step-by-step description of the analytical methodology.

Mapping the Range of OSC and DSC Options

The first step in determining what OSC and DSC systems each side might develop and employ is to identify the range of options available to them. While this is a complex problem, it can be approached systematically. One way to do this would be to develop what might be described as a Master Game Table (MGT), a cross-referenced spreadsheet mapping the full range of potential attacks and counterattacks on space force enhancement systems. To develop such a tool, one would begin by identifying the potential U.S. targets of enemy attack and assessing what kinds of attacks to which they are vulnerable. This would require listing all of the space systems the United States uses to support its warfighting forces, or "Blue space capabilities," and determining how an adversary might attack them, based on the missions they perform, the orbits in which they operate, and the engineering approaches they use to perform their missions. Table 3.1 provides examples of Blue space capabilities and lists some of the orbital regimes in which they operate and some engineering approaches they employ.

Table 3.1. Examples of Blue Space Capabilities, Orbital Regimes, and Engineering Approaches

Mission	Orbital Regime	Engineering Approach
Communications	Low Earth Orbit (LEO), Geosynchronous Earth Orbit (GEO), Highly Elliptical Orbit (HEO)	Radio signal transmission and receipt
Positioning, navigation, and timing (PNT)	Medium Earth Orbit (MEO)	Radio signal transmission
Reconnaissance	LEO, MEO, GEO, HEO	Image (visible, infrared, radar) and radio signal collection
Surveillance	LEO, GEO	Image (visible, infrared, radar), radio signal, and energy signature collection
Environmental	LEO, GEO	Image (visible, infrared) and energy signature collection

Once all Blue space capabilities are identified and characterized by orbital regime and engineering approach, analysts would catalog all conceivable ways an adversary might attack each of those systems, or "Red OSC options." Possibilities emerge from several considerations. A space system consists of three elements: satellites; ground stations; and the links between them, and, in cases of those with broadcast missions, between the satellites and their users. Each of these elements might be attacked in ways that are reversible, such as radio frequency jamming; and the orbital and terrestrial elements may also be attacked in ways that are nonreversible, such as kinetic strikes. Attacks could be non-kinetic (which also might be nonreversible), kinetic, or nuclear. Finally, attacks might originate from the terrestrial domain or from space. Working through all of these possibilities produces a list of potential Red OSC options against each Blue space capability.

The next step is to identify all possible approaches for defending Blue space capabilities, or "Blue DSC options," against each of the Red OSC options arrayed against them. Here, the principal categories to consider are passive defenses (such as shielding and jam-resistant receivers), active defenses (such as kinetic interceptors), and other measures to make the space architecture more resilient, such as adding capacity, rapidly replenishing losses, and dispersing the targeted capabilities across multiple space and terrestrial platforms.[6]

When all Blue DSC options are identified and arrayed against their respective Red OSC options, analysts would then consider what "Red counter-DSC options" might be available to potential enemies. For instance, a possible defense against a jammer targeting a communications satellite downlink could be to install a more powerful transmitter on the next generation of satellites. But that defense would take considerable time to deploy and the opponent might counter it simply by increasing the signal strength of its jammer, thereby reducing the payoff expected from investing in the more powerful space-based transmitter. Therefore, analysts should try to anticipate all possible Red counter-DSC options to help decisionmakers determine whether any particular Blue DSC investment is wise.

The product of this work would be an MGT. Done properly, it would catalog all potential OSC, DSC, and counter-DSC options and discuss their potential benefits and risks in each of the four categories mentioned earlier: mission effectiveness, feasibility, escalation risk, and political cost. Ideally, it would map the universe of conceivable ways

---

[6] Additional capacity, rapid replenishment, and dispersal are not normally classified as DSC actions per se, but it is useful to consider them as such in this analysis so they can be compared with active and passive defenses for mission effectiveness, feasibility, escalation risk, and political cost. In particular, see options to increase the resilience of U.S. space systems listed in Table 2.1.

to attack and defend space capabilities and provide qualitative assessments of each in the aforementioned categories.[7]

Assessing the Options

With all conceivable OSC, DSC, and counter-DSC options identified, analysts can conduct game-theoretic analyses of potential attacks on and defenses of any particular space force enhancement capability by identifying the relevant OSC and DSC options and assessing them for their potential mission effectiveness, feasibility, escalation risk, and political cost. As we shall explain in more detail later in this report, the decision support tool DSPAT is designed to guide users through the game-theoretic methodology, standardizing the assessment process and quantifying selective factors for comparative scoring. In essence, DSPAT leads analysts through each assessment and generates quantitative scores based on the qualitative information they enter in response to prompts provided by the tool. Next, we detail some considerations for assessing OSC and DSC options in each of the following four categories.

Mission Effectiveness

Assessing the potential mission effectiveness of a Red OSC option requires estimating how much its employment would degrade the Blue space capability under attack, *not just the specific element targeted in the attack.*[8] Many factors must be considered in this assessment, such as Blue's vulnerability to that particular type of attack, given the Blue orbit, engineering approach, and how much the targeted element contributes to the overall Blue space capability. For instance, a Red direct-ascent, kinetic ASAT attack on a Blue reconnaissance asset in LEO might degrade Blue reconnaissance capability considerably if the U.S. orbital inventory of that type of satellite is small. However, a comparable attack on a GPS satellite might have a much lower chance of

---

[7] For the purposes of this study, we drafted a working-level MGT with non-program-specific information in it. Because of its limited utility at the level of generality at which it was developed, we decided not to publish this tool. PAF encourages any organization that chooses to employ the game-theoretic methodology described in this report to develop an in-house MGT and populate it with detailed information on space force enhancement programs and prospective OSC and DSC options.

[8] As will be explained in the appendix, DSPAT calculates each OSC option's mission effectiveness score as a product of three factors: (1) the probability that the attacker can successfully execute the kill chain; (2) the percentage of degradation the targeted element will suffer, given a successful attack; and (3) the percentage of overall capability that is lost, given the destruction or complete degradation of that targeted element. There is no standard metric for what constitutes a good or bad mission effectiveness score. Rather, analysts compare mission effectiveness scores across systems to determine which OSC and DSC options the combatants will most likely select.

success, given the altitude of the target, and, if successful, might impose a lesser degradation on Blue's PNT capability, given the robust size of the GPS constellation and the engineering approach it employs.

Assessing the potential mission effectiveness of a Blue DSC option involves estimating how much that defense would reduce the mission effectiveness of the Red OSC attack. Here, some of the factors that analysts must consider are similar to those examined in the OSC case, but some are different. For passive defenses, engineers and other technical specialists must assess how well the proposed defense would protect the Blue targeted element from the effects of the Red attack. For such options as added capacity, rapid replenishment, and dispersal, analysts must determine the degree to which such measures dilute the effects of the OSC attack. In addition, time and cost assessments should be done and weighed against the speed and ease with which Red could absorb the extra capacities with additional attacks. For active defenses, analysts must determine whether the Blue DSC kill chain can be made sufficiently reliable and responsive to detect, characterize, and defeat the Red attack before it manifests against the Blue space capability.

An additional factor that analysts must consider is to what degree employing the Blue DSC option might interfere with the Blue space capability it is trying to defend. For instance, shuttering a sensor might protect its sensitive components from damage from a laser, but it also prohibits that sensor from contributing to Blue capability while the shutter is closed. Similarly, maneuvering a Blue satellite to escape a Red co-orbital interceptor or leave the effect zone of a Red jammer might degrade Blue space capability nearly as much as if the Red attack were successful, should the maneuvers make it impossible for that satellite to contribute to the mission in the short term, or should they force it to exhaust its propellant, shortening its life in the long term.

Considerations made when assessing the potential mission effectiveness of Red counter-DSC options are similar to those required for Blue DSC options. However, Red has some inherent advantages. First, Red has a first-move advantage in both OSC and counter-DSC attack execution. Whereas Blue must detect and characterize the Red attack before it can formulate and execute a response, Red knows what Blue capability it is attacking and can anticipate the DSC options Blue might attempt to employ. Second, whereas Blue must defend itself in a way that preserves the space capability it is defending, the sole mission of most prospective OSC capabilities is attacking Blue. Therefore, Red can afford to expend all of its energy or propellant—or even act in a way that destroys its weapon—if that action results in the destruction or severe degradation of the targeted element of the Blue space capability.

Feasibility

Feasibility is the assessment of whether a given OSC or DSC will be operationally available to its prospective user by some specified date for which a particular analysis is being conducted. For instance, if analysts are assessing possible Blue DSC options against potential attacks by a specific Red actor on a Blue space capability in 2020, they would review intelligence threat assessments to determine which of the universe of conceivable OSC options that actor would likely have available for use by that date. Those options that threat assessments indicate with high confidence Red will have in its operational inventory should be included in the analysis. Those options that the Intelligence Community assesses a low probability of operational availability by that time should be omitted. When threat assessments are uncertain or expressed with medium confidence, analysts may choose to run analyses with and without those Red OSC options to determine the degree to which options might threaten Blue space capabilities and affect escalation dynamics should they appear and be employed.

Similarly, analysts would consult U.S. space system developers, national laboratories, and other sources to determine which Blue DSC capabilities are on track for availability in 2020 and what additional capabilities might possibly be developed, given increases in effort and funding, should analysis indicate a justifiable need for those capabilities. Systems that will likely be available by the target date should be included in the analysis. Analyses should be conducted with and without systems that could be available with additional funding to help determine whether the added capability they provide justifies the resources required to rush them to operational status.

In both sides of the analysis—i.e., that of OSC options and DSC options—the feasibility assessment operates as a filter: Those options deemed feasible by the year in question are retained in the set of candidates for further analysis; those deemed infeasible by that year are discarded.

Escalation Risk

In the context of international confrontation and war, the word *escalation* refers to any increase in the intensity or scope of conflict. Increases in the intensity of attacks are often described as *vertical escalation* and are related to rises in the frequency of attacks or the levels of destruction they cause. Increases in the scope of conflict are described as *horizontal escalation* and involve spreading a conflict to areas, states, regions, or domains not previously affected by it. Belligerents typically escalate conflicts vertically or horizontally to gain advantage or avoid defeat. They may also do so to signal resolve or achieve some coercive objective, such as compelling an adversary to back down. But explaining escalation solely in terms of the foregoing motives would suggest that escalatory acts are always deliberate. Unfortunately, they are not. Escalation can also be inadvertent, when one belligerent does something he does not consider escalatory, but the enemy does and escalates in response, and escalatory acts can be accidental, when one's forces make mistakes, such as striking the wrong targets.[9]

Assessing what escalation dynamics might result from the employment of alternative OSC and DSC options is somewhat speculative, as belligerents have done little to attack space capabilities in previous conflicts and we therefore have almost no historical data to evaluate. However, we can estimate how escalatory specific actions will appear based on the intensity of those attacks, where they manifest, and whether, during a conflict in question, attacks of comparable intensity have previously occurred in those regions. As Figure 3.3 illustrates, these considerations result in a two-dimensional graph on which we can plot the approximate levels of vertical and horizontal escalation that OSC and DSC actions have reached at any given point in a notional conflict.[10]

Figure 3.3. Qualitative Escalation Risk Matrix

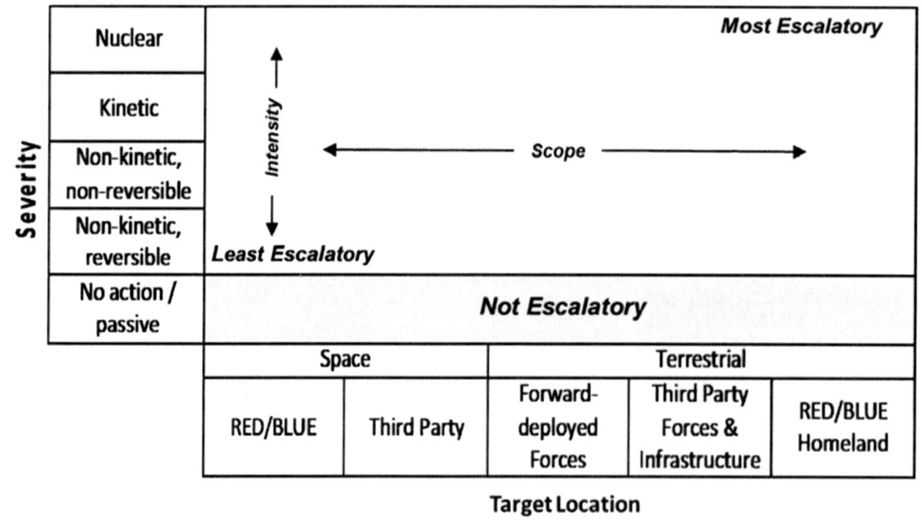

As shown here, taking no action and simply relying on passive defenses is never escalatory. A non-kinetic, reversible-effects attack is escalatory if none have yet occurred

---

[9] For more on the nature of escalation and approaches for managing it, see Forrest E. Morgan, Karl P. Mueller, Evan S. Medeiros, Kevin L. Pollpeter, and Roger Cliff, *Dangerous Thresholds: Managing Escalation in the 21st Century*, Santa Monica, Calif.: RAND Corporation, MG-614-AF, 2008.

[10] As will be explained in the appendix, DSPAT elicits the information shown in Figure 3.3. It then plots each OSC and DSC option on the Quantitative Escalation Risk Matrix (shown in Table A.1) and assigns a numeric baseline escalation score.

in a confrontation or conflict. Non-kinetic, non-reversible-effects attacks, kinetic attacks, and nuclear attacks are increasingly escalatory in turn. How escalatory any of these events are, however, also depends on where they occur. We generally consider attacks on terrestrial targets more escalatory than those aimed at orbital assets, because terrestrial attacks are more likely to take human life, but in both domains, much depends on what assets are harmed and whose lives are put at risk.[11] Attacks in space that threaten third-party assets are more escalatory than those that only harm satellites owned by the belligerents, even if the third-party harm is unintended. Similarly, attacks in the terrestrial domain that harm third-party forces, citizens, or infrastructure are more escalatory than those that discriminately target an enemy's opposing forces. Attacks on a belligerent's homeland are more escalatory than those on its forces abroad.

It is important to remember, however, that whether any given act is escalatory depends on whether the other side has already breached that threshold. Escalation is conditional. For instance, should a belligerent employ a space-based jammer against an enemy communications satellite, that OSC option would be escalatory if neither side had previously employed a reversible-effects attack in space during that conflict. The defender might respond by maneuvering away from the jammer by relying on a passive defense, such as jam-resistant transceivers, or even by jamming the uplink to the attacking satellite, none of which would be escalatory, because the OSC attack just employed had already crossed the reversible-effects attack threshold in the space domain. However, should the defender respond by attacking the jammer with a kinetic weapon, that would escalate the conflict further, and he should not be surprised if his orbital assets are thereafter targeted for kinetic attacks. Alternatively, should the defender respond by bombing the ground station that controls the jammer, that act would be escalatory if the ground station were outside the region of terrestrial conflict, but not escalatory if the conflict had already engulfed that area and conventional strikes had commenced on other targets there, whether or not they were space-related.

Finally, when comparing a potential opponent's space control options for escalation risk, it is important to remember that different states have different tolerances for escalation, and those tolerances vary depending on the levels of interest at stake in the conflict. For instance, should China find itself in a military confrontation with Japan over the Senkakus, Chinese leaders would likely have a relatively low tolerance for escalation risk and would choose their space control options accordingly. Alternatively, were North

---

[11] An exception to this principle exists regarding reversible-effects attacks, which might be more escalatory when done against certain orbital assets than when they are done against terrestrial forces. This exception is not shown in Figure 3.3, but we capture it in the quantitative scoring in DSPAT, which is illustrated in Table A.1, Quantitative Escalation Risk Matrix, in the appendix to this report.

Korea to become engaged in a major conventional war with South Korea and the United States, Pyongyang might be much more willing to escalate the fight in space if North Korean leaders believed doing so would increase their mission effectiveness in ways that better assured the successful accomplishment of their military and political objectives. Game-theoretic analyses of both of these cases could be done using the same escalation risk matrix—for instance, both states would likely consider kinetic attacks more escalatory than non-kinetic attacks, nuclear attacks more escalatory than conventional attacks, and they would agree that attacks that harm third parties would risk widening a conflict—but a state fearing for its survival would be willing to bear much greater escalation risk than one fighting for more peripheral interests. Analysts will need to take these considerations into account when anticipating space control options any particular opponent might choose in a given scenario.

Political Cost

In the context of assessing alternative OSC and DSC approaches, *political costs* are defined as those a belligerent might incur in the censure of third parties, domestic and international, as a result of its behavior. Political costs can manifest in a variety of ways, depending on how objectionable global audiences perceive the offense to be. Costs may be as mild as expressions of disapproval from domestic political opponents or as severe as widespread international condemnation, economic sanction, or expulsion from intergovernmental organizations.

All states are sensitive to political costs, although to varying degrees depending on their particular circumstances. Any state fighting for its survival will suffer whatever censure it must to defend itself, but most states in confrontations short of war and even in limited conflicts will act in ways that mitigate their political costs, if possible, to keep them within acceptable limits vis-à-vis the expected benefits of their actions. How sensitive a state is to political costs depends on how responsive its leaders are to domestic constituents, how much it values its prestige in the community of nations, and what benefits it hopes to reap from other states' cooperation or support. The United States and other liberal democracies—nations that cherish their reputations as responsible global actors and whose leaders depend on popular support—are the most motivated to avoid visible actions that might tarnish their images. At the other extreme, such states as North Korea, with cowed domestic audiences and little to lose in national prestige, are the least concerned about what third parties think of them. Even Pyongyang, however, must weigh its actions in how it might affect future prospects for international assistance and relationships with its few friends, particularly China.

It is impossible to predict what political costs any state, even our own, would be willing to bear at any particular level of conflict. However, when evaluating OSC and

DSC options, it is important to compare them for their potential political costs because decisionmakers will weigh those costs, along with others, against benefits they expect the prospective systems to create. If two alternative DSC approaches offer comparable levels of mission effectiveness and escalation risk, but one risks high political costs, political leaders would likely prefer to develop and employ the less costly option, particularly if they expect an opponent to employ the OSC option it is designed to defeat at low levels of conflict.

Assessing an OSC or DSC option's risk of political cost requires analysts to consider multiple factors. First, would the employment of that option be observable to third parties? As in calculating escalation risks, passive defenses bear no political cost. Similarly, attacks that cannot be detected—i.e., those mistaken for equipment failures or degradations caused by natural phenomena—generate no political costs. Even if the targeted belligerent knows it has been attacked, the attacker's political costs will be low if third parties can see no evidence of it, because enemy leaders will simply deny accusations that they perpetrated an attack and many third parties will believe them. If, on the other hand, the OSC or DSC option or its effects can be seen by others, the question becomes: To what degree is it attributable to the perpetrator? Acts that might be done anonymously, such as the sowing of space mines, risk less political cost than overt attacks, such as the use of a direct-ascent ASAT from one's home territory. Yet, even anonymous attacks risk some political cost, because third parties will likely consider it suspicious when something occurs in space during a crisis or war that damages their assets or someone else's.

Assuming an OSC or DSC option is observable and attributable to some extent, analysts must assess a range of other factors to determine the degree of political cost the act might incur. These factors include where the attack would manifest, what property or people would be harmed, and how severe that harm would be. The assessment must also consider whether the attack would violate certain international norms, such as taboos against nuclear weapons use or warfare in space.[12]

---

[12] Although international treaties prohibit detonating nuclear weapons in space, placing weapons of mass destruction in orbit or on celestial bodies, and stationing military forces on celestial bodies, no treaty prohibits placing conventional weapons in space, using them against space assets, or using them against terrestrial targets from space. That said, idealistic statements made in treaties and government pronouncements about space being the endowment of all humankind, coupled with the fact that states have largely avoided conducting warfare in or from orbit for the first five decades of the space age, have led many people to believe that space has become a sanctuary from war and should be maintained as such. As a result, a taboo against space warfare has emerged, and states that violate that taboo will likely face censure from domestic and international audiences. For informed discussions on this issue, see Colin Gray, "Space Arms Control: A Skeptical View," *Air University Review*, November–December 1985; David W. Zeigler, "Safe Heavens: Military Strategy and Space Sanctuary," in Bruce M. Deblois, ed., *Beyond the Paths of*

Given these many considerations, estimating political costs across the broad range of possible space control activities is a complex endeavor.[13] For instance, reversible-effects attacks in the terrestrial environment would not likely incur political costs. Should Red or Blue conduct those attacks against assets in space, however, either would likely prompt some amount of international criticism for violating the space warfare taboo. If assets belonging to a third-party state, i.e., "Green" assets, were harmed, the costs would be greater, especially if the attacks were to cause Green casualties. Non-kinetic, non-reversible-effects attacks—i.e., those that cause permanent damage—are more costly, in both the terrestrial environment and space, and kinetic attacks are costlier still. Reversible-effects kinetic attacks in space, such as those in which astronauts or robotic devices temporarily interfere with another state's satellites, might be somewhat less costly, but it could be difficult for the victim or third parties to determine whether the effects were only temporary until they were reversed. Kinetic attacks that generate space debris would be considerably more costly than those that do not, as the debris from those attacks would indiscriminately endanger Green orbital assets. Those attacks would be even more costly if they targeted Green satellites, with the debris endangering even more Green assets.

The OSC and DSC options with the highest political cost would involve the use of nuclear weapons. Such attacks would not only violate strong international taboos, they would cause indiscriminate collateral damage and casualties. The probability of noncombatant casualties would be greatest in the terrestrial environment, but a nuclear detonation in space could also endanger noncombatant lives if any people are in orbital vehicles or stations at the time of the attack.

In all of these assessments the general principle is this: The more a Red or Blue action visibly endangers or harms noncombatant property or lives or violates international

---

Heaven: The Emergence of Space Power Thought, Maxwell Air Force Base (AFB), Ala.: Air University Press, 2003, pp. 185–245; Michael Krepon, Theresa Hitchens, and Michael Katz-Hyman, "Preserving Freedom of Action in Space: Realizing the Potential and Limits of U.S. Spacepower," in Charles D. Lutes and Peter L. Hays, eds., *Toward a Theory of Spacepower: Selected Essays*, Washington, D.C.: National Defense University Press, 2012, pp. 119–136; and Bruce M. Deblois, "Space Sanctuary: A Viable National Strategy," *Airpower Journal*, Winter 1998. For arguments that the United States should place weapons in space despite political costs, see Everett Dolman, *Astropolitik: Classical Geopolitics in the Space Age*, London: Frank Cass, 2002.

13 Due to the many considerations involved, political costs do not ascend in a consistently linear manner when the factors are arranged in a two-dimensional table, such as the one shown in Figure 3.3, where we illustrated qualitative escalation risks. Therefore, we do not provide a comparable qualitative table of political costs. However, we have modeled these factors in DSPAT and provide the formula for calculating political costs, along with the table it draws values from, in the appendix to this report. See appendix section "Scoring Political Cost."

taboos, the more costly it will be for the perpetrator's domestic and international relations.

How the DSPAT Supports the Game-Theoretic Methodology

This chapter introduced the game-theoretic methodology and described the conceptual development and application of an MGT. The MGT would be a valuable tool, but its development would be a lengthy project that might take more time and resources than many organizations could devote to it. Moreover, its size and complexity might render it too cumbersome a device for easy use in more extensive game-theoretic analyses. Therefore, to streamline the analysis and make it feasible without the support of an MGT, we have developed the DSPAT to lead analysts through the combinatorial game process and automate the scoring of each OSC and DSC option examined. Figure 3.4 shows how the DSPAT works to provide information for decisionmakers.

Figure 3.4. Game-Theoretic Methodology Inputs and Outputs

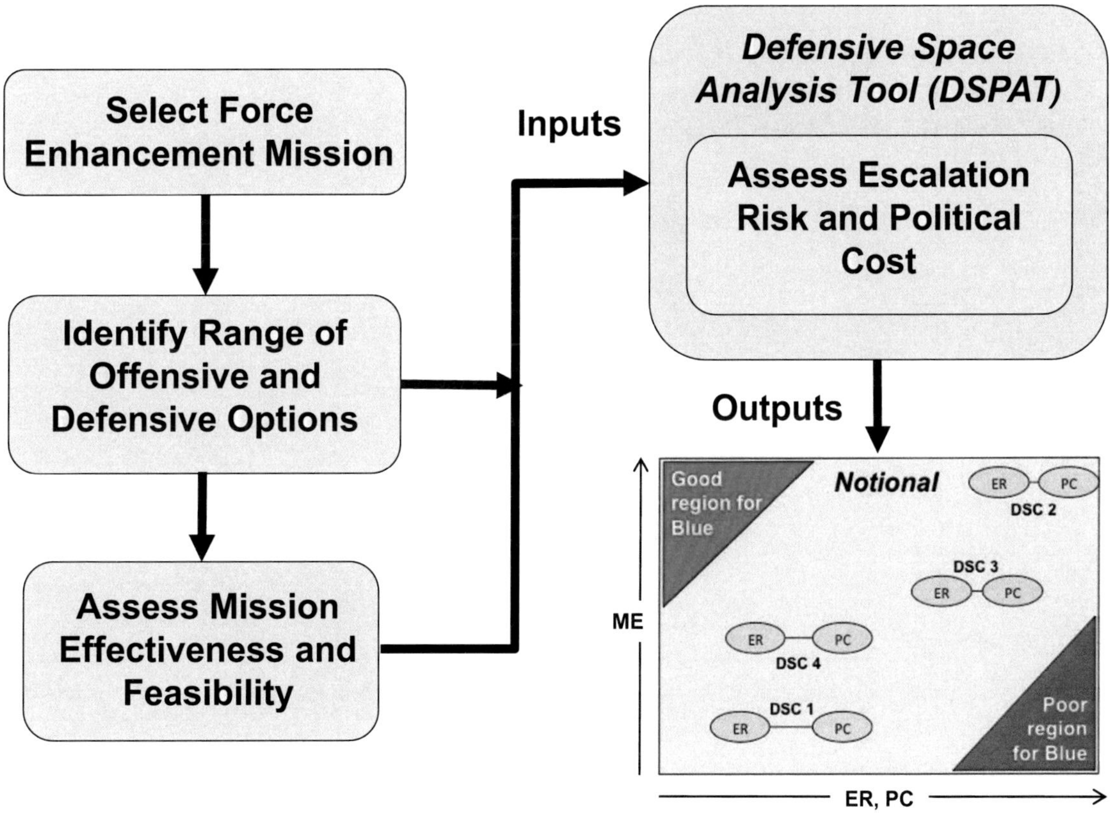

As previously explained, the DSPAT is a decision support tool that solicits qualitative and quantitative inputs from subject-matter experts (SMEs), assigns relative values to the qualitative information it receives, and then submits those values to an integrated set of mathematical models that automate the scoring and ranking process in each of the four assessment categories. To begin a game-theoretic analysis, analysts first select a space force enhancement mission area to evaluate and identify some range of ways to attack that capability, or "offensive options," and a range of ways to defend against each of those attacks, or "defensive options." Then, they assess the mission effectiveness and feasibility of each of the offensive and defensive options under investigation and enter that information into the DSPAT. The DSPAT scores each option for escalation risk and political cost and ranks the outputs for decisionmaker review.

Figure 3.5. Display of Notional DSC Analysis Outputs

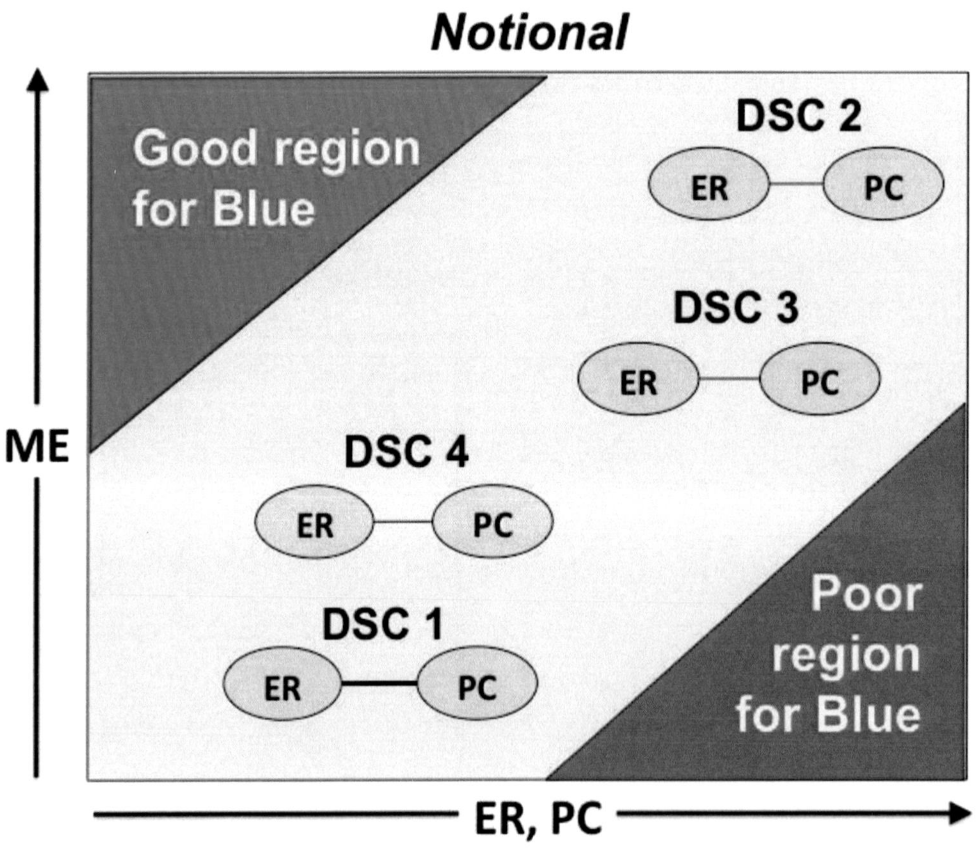

A notional outcome of the methodology is shown in the lower-right corner of Figure 3.4, and this illustration is expanded in Figure 3.5. In this example, four Blue DSC options were evaluated as possible ways to protect a Blue space capability from a specified Red OSC threat. Mission effectiveness (the ability to counter the Red OSC and restore Blue space capability) is plotted along the y-axis and escalation risk and political cost are plotted along the x-axis. Thus, placement in the upper-left corner would indicate high mission effectiveness and low escalation risk and political cost—a good option for Blue. Placement in the lower-right corner would indicate low mission effectiveness and high escalation risk and political cost—a poor option for Blue. In this example, DSC-2 represents a saddle point in which the option that scored highest in mission effectiveness also scored highest in escalation risk and political cost. Had the options clustered in the middle of the graph, it would have indicated that the analysis ended in an indeterminate draw.

The next step in understanding the game-theoretic methodology is to examine its application in various modes of analysis. We embark on that task in Chapter Four.

4. Types of Analysis

This chapter describes four types of analysis in which the game-theoretic methodology can be applied through a variety of space deterrence and escalation management-related questions. In the first type, analysts apply the process in a *Dynamic Strategic Assessment* to identify which OSC and DSC options would likely be most attractive to belligerents on each side of a confrontation, thereby anticipating potential paths of conflict and escalation in space. The second consists of a *Threat-Driven Assessment* in which analysts evaluate a particular OSC option to determine the most effective DSC counters to it within acceptable levels of escalation risk and political cost. The third type of analysis consists of a *Technology-Driven Assessment* of particular DSC options that U.S. leaders might be contemplating to evaluate their merits and determine whether opponents might be able to easily counter those defenses. Finally, the analytical process can also be used in reverse to do a *Blue OSC Assessment* to identify the ways to attack enemy space capabilities with the highest effectiveness, least escalation risk, and lowest political cost.[1]

Dynamic Strategic Assessment

The type of analysis that most broadly applies the game-theoretic methodology can be described as a Dynamic Strategic Assessment. In this type, illustrated in Figure 4.1, analysts choose a specific Blue space capability and evaluate the full range of ways to attack and defend it.

---

[1] Readers who use DSPAT in support of the game-theoretic assessments described in this report will notice that the DSPAT input panels and *DSPAT User's Manual* refer to three "modes" of analysis, versus four "types" of analysis. DSPAT's modes of analysis correspond to the types of analysis described here. Only three are addressed in DSPAT and the *User's Manual* because the second mode, the Selected OSC Mode of Analysis, is used in two types of analysis described here: the Red Threat-Driven Assessment and the Blue OSC Assessment. See Forrest E. Morgan, James Syme, and Christopher Lynch, *DSPAT User's Manual*, Santa Monica, Calif.: RAND Corporation, TL-121-AF, 2017.

Figure 4.1. Dynamic Strategic Assessment

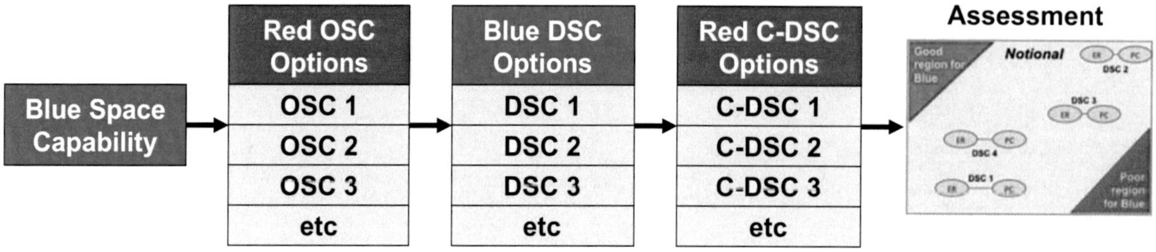

This type of analysis most thoroughly captures all of the steps, considerations, and techniques described in Chapter Three. It begins with a mission effectiveness–driven ranking for the list of potential Red OSC options. Analysts then filter OSC options by feasibility and rank them by escalation risk and political costs. Once all OSC options are evaluated and ranked, analysts consider the scores and select the Red OSC option they conclude would be most attractive to the specific Red actor being considered in the analysis, given its need to achieve a space control mission and its tolerances for political costs and escalation risk at whatever level of conflict under consideration.

With the most likely Red OSC option identified, analysts then list all potential approaches for defending the Blue space capability from that particular attack and evaluate how well each Blue DSC option reduces the effectiveness of the Red OSC option it is meant to defend against. They also consider how long it would take each option to take effect, comparing it with the time-to-effect assessed for the selected OSC attack, and how much, if any, employing each DSC option would degrade Blue mission capability. They further consider whether each DSC option would make the Red OSC option it is defending against more observable to Green and attributable to Red, adjusting Red OSC political cost scores accordingly. Analysts then filter the DSC options by feasibility and rank them by escalation risk and political cost.

When all relevant Blue DSC options are evaluated and ranked, analysts choose the most effective one with acceptable escalation risks and political costs for Red counter-DSC option assessment. As explained in Chapter Three, if an effective counter is identified, the analysis cycles back to the Blue DSC list and Blue Option 2 is selected. If Red has no effective counter, it cycles back to the Red OSC option list and Red Option 2 is selected. Blue DSC options are then assessed against that attack and Red counter-DSC options in turn (see Figure 3.2).

Ultimately, iterative analyses will either reveal a dominant strategy or saddle point, or result in an indeterminate draw.[2] In any of these outcomes, analysts can review the overall analysis and assess which paths of conflict would likely emerge in a fight to attack and defend this particular space capability, what levels of escalation would result from it, and what political costs each side would likely accrue as a result.

Analysts could conclude the process at this point and prepare an assessment report or briefing for decisionmakers. Alternatively, they could continue iterative analyses until all Red OSC options are run against all Blue DSC options and all of those are examined against all Red counter-DSC options. With an exhaustive analysis completed on one Blue space capability, analysts could then proceed to the next, and so on, until all OSC, DSC, and counter-DSC matchups are analyzed across all Blue space force enhancement capabilities. A comprehensive analysis of this kind would reveal the most frequently predicted paths of conflict across the universe of possibilities, illuminating the levels of escalation and political cost that Blue and Red decisionmakers should expect should conflict occur in space.

Red Threat-Driven Assessment

While doing a full Dynamic Strategic Assessment would yield a great deal of useful information, decisionmakers often want more-focused analyses. For instance, should intelligence reports indicate that a Red actor appears committed to developing a specific OSC capability, the game-theoretic methodology could be applied in an analysis specifically tailored to evaluate the implications of such a development and identify the most-desirable approach for defending against it. Figure 4.2 illustrates the *Red Threat-Driven Assessment.*

---

[2] We discuss the implications of each of these outcomes in Chapter Five, "Interpreting the Outcomes of Game-Theoretic Analyses."

Figure 4.2. Red Threat-Driven Assessment

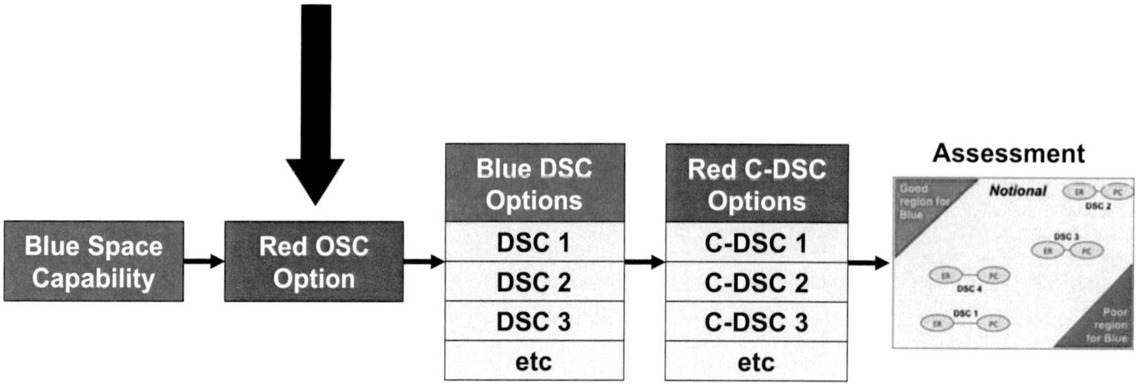

This type of game-theoretic analysis is simpler and more direct than the strategic dynamic assessment. In this case, analysts not only know what Blue space capability is being threatened, but also know how Red intends to attack it. Therefore, they can omit the analysis of multiple OSC options and, instead, do a focused analysis of the specified threat to establish baseline scores in mission effectiveness, escalation risk, and political costs. Analysts then identify and list all potential ways of defending against that particular Red OSC option. From there the analysis proceeds as explained above. All listed DSC options are evaluated and ranked by mission effectiveness. Those that offer the highest levels of mission effectiveness and are deemed feasible are ranked according to escalation risk and political cost. The most-desirable defense with acceptable levels of escalation risk and political cost is then subjected to Red counter-DSC evaluation. If Red cannot counter the DSC, the effective defense to Red's OSC has been identified and should be developed. If Red can counter it, analysts cycle back to the ranked Blue DSC list and select the next most-desirable option, cycling between the Blue DSC options list and Red counter-DSC options until a defense is found or the list is exhausted. If the latter occurs, U.S. leaders must be informed that Red has discovered a dominant space control strategy: It is developing a system that is threatening Blue space capabilities in a way that cannot be defeated. Implications of this finding are discussed in Chapter Five.

Blue Technology-Driven Assessment

From time to time, space program offices or system developers propose new systems for defending U.S. space capabilities or enhancements for making systems more robust against attack. Sometimes these proposals entail developing active defenses; sometimes they involve developing passive defenses or making architectures more resilient by adding redundancy or dispersing key elements. Whenever such ideas are discussed, questions often arise as to how effective the proposed technology would be against which threats and the degree to which developing or employing it would be escalatory or politically costly.

Such questions can be addressed using the type of game-theoretic analysis shown in Figure 4.3, the *Blue Technology-Driven Assessment*.

Figure 4.3. Blue Technology-Driven Assessment

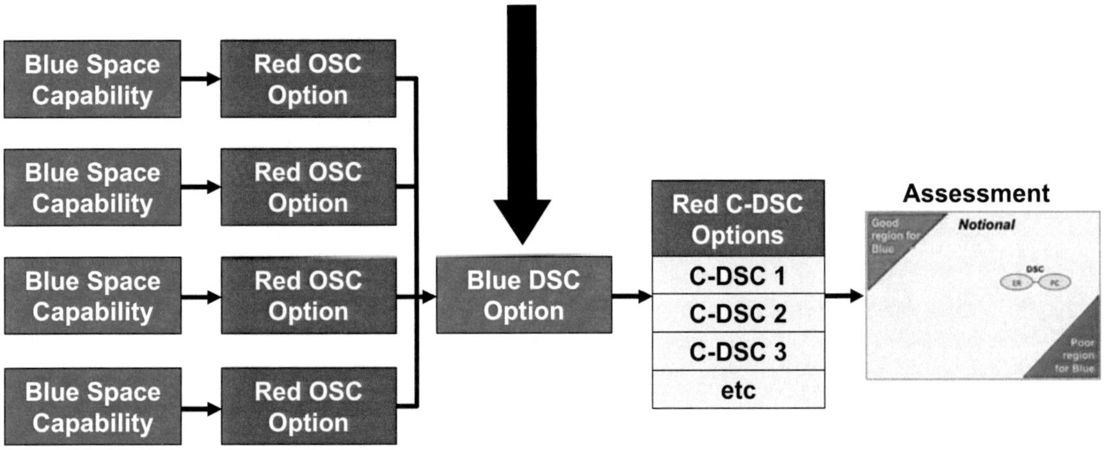

In this assessment, analysts identify the Red OSC option that the proposed Blue technology is presumed to defend against and evaluate its baseline mission effectiveness, feasibility, escalation risk, and political cost as described earlier in the Red Threat-Driven Assessment. With baseline scores for the Red OSC option thus established, analysts then assess the proposed Blue DSC option's mission effectiveness against that OSC attack and its feasibility, escalation risk, and political costs. They also conduct an assessment of potential Red counter-DSC options, as explained in the earlier sections on Dynamic Strategic and Red Threat-Driven Assessments.

By this point in the process, analysts will know how effective the proposed DSC option would likely be against the OSC capability that prompted calls for its development, and they will also understand the escalation risks and political costs it could generate. This knowledge alone might be enough to justify or cancel a program. However, another question decisionmakers will likely ask is whether the proposed DSC approach would be effective in defending against other threats in addition to the one that prompted its consideration. As Figure 4.3 illustrates, a system that protects against four dangerous threats to U.S. space capabilities is more justifiable than one that defends against only a single threat, especially if the proposed system would be costly to develop.

To answer this question, analysts could turn to the MGT, if they have developed one, and review all OSC approaches listed which attack Blue space capabilities in ways similar to the OSC option first assessed and survey the Blue DSC options listed for each of them. Assuming analysts have, in their MGT development, attempted to map the universe of attacks on and defenses of Red and Blue space capabilities, other applications of the proposed Blue DSC should appear in those lists, if they are plausible. Analysts should identify them and then assess the proposed DSC approach against the Red OSC options identified in each of those cases. If no MGT has been developed, analysts can identify all additional Blue space force enhancement systems that the Red OSC capability in question might threaten and, taking the orbital parameters and engineering principles of each system into consideration, also evaluate the degree to which the proposed DSC could defend those systems. Decisionmakers can then be apprised of the full range of defenses that the DSC option under consideration could provide, giving them a more accurate appraisal of its value.

Blue OSC Assessment

Although U.S. leaders are generally more interested in defending U.S. space capabilities than attacking those of potential adversaries, there are situations in which it might be advantageous to have certain Blue OSC capabilities available to posture in peacetime for deterrent purposes or employ in war for space control. But which Blue OSC capabilities would offer the highest mission effectiveness with manageable levels of escalation risk and acceptable political costs?

As Figure 4.4 illustrates, the game-theoretic analysis can be applied in a Blue OSC assessment to answer this question.

Figure 4.4. Blue OSC Assessment

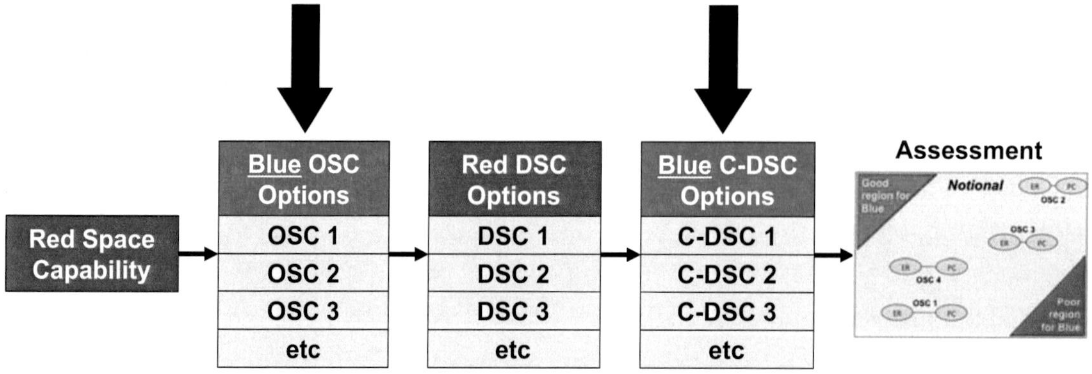

Whether analysts choose to develop an MGT, they will probably organize the range of moves and countermoves they are assessing under the assumption that the United States (Blue) will be defending its space capabilities against an opponent's (Red) attacks. However, the analysis would be equally valid with the colors reversed. That is, the analysis could also map some range of possible Blue attacks against Red space capabilities, Red defenses against those attacks, and Blue counters to those defenses.[3]

To conduct a Blue OSC assessment, analysts would determine what Red space capability U.S. leaders might want to attack. They then would develop a list of OSC options that could be used to attack that space capability and assess them for mission effectiveness, feasibility, escalation risk, and political cost. When the analysis identifies what appears to be the best OSC option for attacking the Red space capability, analysts could then conduct DSC assessments against it and counter-DSC assessments against the best DSC option that emerges, cycling through iterations of analysis, as described in the foregoing sections, until the best Blue OSC option is identified.

While an MGT could be used in this fashion simply by reversing the colors, it is important to point out that, whether an MGT is used or the analysis is done with DSPAT alone, analysts should not take the findings of a previously run Red OSC analysis in DSPAT and simply reverse the colors, assuming that it has also identified the best Blue OSC option. Red actors have space capabilities that differ in quantity and quality from those the United States operates, and the percentages of force enhancement support they get from space versus terrestrial systems differs from what U.S. forces receive. Therefore, game-theoretic analyses will produce different outcomes for different actors, and analysts will need to do Blue OSC assessments with specific Red opponents in mind.

Summary

This chapter explains how analysts can apply the game-theoretic methodology in four types of OSC and DSC analysis. With the process thus explained, it is time to turn our attention to interpreting the outcomes of game-theoretic analysis. We undertake that task in Chapter Five.

---

[3] DSPAT has modules designed to guide assessments of Blue OSC options against Red space capabilities and the subsequent analyses of Red DSC options and Blue counter-DSC options.

5. Interpreting the Outcomes of Game-Theoretic Analyses

This chapter explains how to interpret the outcomes of analyses using the sequential game-theoretic methodology. As previously mentioned, each type of analysis described in Chapter Four would result in one of several outcomes: the emergence of a dominant strategy for one side or the other; a saddle point in which one side or the other would have effective OSC or DSC options at its disposal, but with very high escalation risks, political costs, or both; or an indeterminate draw in which both sides have some number of space control options available, none of which is particularly effective. Here, we discuss the implications of each of these outcomes for deterrence and dissuasion, escalation dynamics, and investment decisions concerning the development of alternative space control capabilities.

Implications of a Dominant Strategy

The possibility that a dominant strategy will emerge for one side or the other exists in all applications of the game-theoretic methodology. That condition exists when an actor discovers an attack or defense that achieves its mission requirements with acceptable levels of escalation risk and political costs and for which the other side either has no feasible, affordable, effective counter, or none is available with acceptable levels of escalation risk and political cost. Should analysts discover such an approach for defending an important U.S. space capability, they should advocate its development.[1]

Whether such a system, once developed and postured, would deter adversaries from attacking the defended capability would depend on the nature of the DSC approach in question. If the DSC capability is sufficiently powerful and robust that it can defeat all known OSC options, even if opponents know it exists, then U.S. leaders should deploy it and make potential adversaries aware of its existence. Knowledge that the United States has a defense that can defeat all OSC attacks against a particular U.S. space capability should deter potential adversaries from attacking that capability, as the attack would not yield them any benefit and would likely result in the United States punishing them in some way.[2] Should the most promising OSC option against that capability be expensive

---

[1] This assumes, of course, that the system in question can be developed and fielded within cost constraints, which is part of the feasibility assessment.

[2] Punishments for attacking U.S. space capabilities need not consist merely of tit-for-tat strikes against opponent space assets, or even military actions alone. U.S. leaders should consider a wide range of

to develop, the opponent might even be dissuaded from investing the resources in attempting to do so, given the expectation that it would be defeated anyway. Even if the defensive capability cannot decisively defeat all known OSC approaches, one that can defeat some of Red's most attractive options or substantially degrade their effectiveness could change a potential opponent's cost-benefit analysis considerably, thus strengthening deterrence, while providing some defense should deterrence fail.

Some defensive capabilities, however, may not contribute to deterrence even if they are effective. Those capabilities whose effectiveness depends on keeping their existence secret cannot be the substance of specific deterrent threats, because to threaten their use would reveal their presence, prompting opponents to take measures to defeat them. That does not suggest that such capabilities should not be developed. If they offer effective and affordable defenses, U.S. leaders should have them available for use in the event of war. Moreover, it might be possible to fashion deterrence strategies around some of these capabilities by signaling their existence in general ways without revealing the technical details or specific locations that would allow an opponent to counter them. Careful analysis should be done on each of these capabilities to determine what messaging strategies would best support deterrence without undermining defense.

Finally, analysts and decisionmakers must understand that the discovery of a dominant capability is never the endpoint of strategic analysis. The dominance of any military technology tends to be a temporary condition. The fact that the United States has developed a system that effectively defeats all threats to an important space force enhancement capability will inevitably inspire potential adversaries to discover ways to defeat that system or develop new ways to threaten the defended capability. Analysts will need to rerun game-theoretic assessments to determine the implications of each new threat as it emerges and even new concepts from imaginative system developers and strategy professionals.

Dealing with Saddle Points

Another possible outcome of the space control game-theoretic analysis is that a saddle point will emerge. In that condition, one side or the other can acquire a space control capability with a high level of mission effectiveness, but the capability would bring with it high levels of escalation risk, political cost, or both. For instance, one could consider

---

potential ways to retaliate against opponents that attack U.S., allied, and third-party space capabilities—diplomatic, economic, and informational, as well as military—and they should threaten to exact retribution in times and places of U.S. choosing. For a discussion of how to make U.S. deterrent threats of punishment more potent and credible, see Morgan, 2010, pp. 38–44.

China's direct-ascent, kinetic ASAT capability to be a saddle point strategy against U.S. reconnaissance capabilities in LEO, were there no effective defenses against it.

Saddle point outcomes have particular implications for U.S. strategists, system developers, and decisionmakers. Given high levels of mission effectiveness coupled with high escalation risks, they suggest strong escalatory pressures could emerge around these options during a crisis or conflict, especially if the space force enhancement capability at stake is particularly threatening to the attacker and valuable to the defender. If intelligence threat assessments indicate that a potential adversary is developing and posturing OSC capabilities with saddle point characteristics, it implies one or both of two possible motives: First, the Blue space capability that the OSC system would target is sufficiently threatening to that Red actor that it is willing to risk high levels of escalation and political cost in hopes of deterring Blue from engaging in war with it. Second, if war occurs, enemy leaders are willing to escalate in space, or they expect the scope or intensity of conflict to otherwise rise to a level at which employment of the OSC capability would no longer be escalatory. It also suggests they would be willing to pay high political costs to win.

Should the Intelligence Community determine that such a system is being pursued, U.S. leaders should devote considerable efforts to developing a counter to it. The opponent's commitment to fielding a high-risk, politically costly system indicates a strong desire to hold U.S. capabilities at risk. Ideally, U.S. developers can find ways to defend against the emerging threat that are low in escalation risk and political costs, while raising the opponent's potential political costs by increasing third-party awareness of the threat and their ability to attribute it to Red. However, even if the only options the United States can develop for countering the Red threat would be escalatory and politically costly, U.S. leaders may want to develop them anyway. The potential opponent's commitment to developing the ability to conduct such a provocative attack indicates its leaders expect a conflict with the United States to escalate to levels at which such weapons might be used. U.S. leaders should, therefore, be prepared to fight and win at those levels as well.

Decisions on whether the United States should develop DSC or OSC capabilities on which game-theoretic analyses result in saddle point outcomes hinge on several factors. First is the number and criticality of the space force enhancement capabilities the DSC option could defend or the OSC option could attack. Options that would enable defenses of, or attacks on, multiple, important force enhancement capabilities might be worth developing despite their escalation risks or political costs. But DSC options of this nature would likely be more attractive to U.S. leaders than OSC options, because U.S. forces are much more heavily dependent on space force enhancement than are any of their potential adversaries. U.S. leaders would likely be willing to risk escalation or pay political costs in defense of their own critically important space capabilities than to attack the less important capabilities of opponents. However, much depends on the specific systems the Blue OSC capability would target and how Red might intend to use those systems to enhance its warfighting effectiveness in specific scenarios.

The second consideration is at what level of conflict the DSC or OSC option would likely be employed. DSC options that, considered in isolation, would be escalatory in crises or low-level conflicts would not be escalatory if used to defend against attacks that would have already crossed the same escalation threshold.[3] Saddle point OSC options are less likely to benefit from a threshold being previously breached; however, they also are not escalatory if they are withheld until the conflict has already escalated beyond the thresholds they would have crossed. In either case, were any of those DSC or OSC options politically costly, the United States would pay penalties for their use regardless of whether an adversary had used them first.

A third consideration for saddle point DSC options is whether they would be sufficiently effective to deter attacks on the defended space capabilities. Developing a highly effective defense, even one that would be escalatory or politically costly if employed, might dissuade an opponent from developing certain OSC options or deter it from employing them even if available in a crisis. The United States' commitment to defend its space capabilities is one that potential adversaries cannot take lightly, and openly developing DSC capabilities that are escalatory or politically provocative would underscore that commitment. That said, if comparable mission effectiveness could be achieved with DSC options that are less escalatory or politically costly, U.S. leaders would likely prefer to develop those instead, and threats to employ them would be more credible.

The Indeterminate Draw—A Glass Half Full

A less satisfying outcome of space control game-theoretic analysis is the indeterminate draw, when neither side has a dominant strategy, and both have one or more options that offer mediocre results. While such outcomes are discouraging, they are informative and not necessarily detrimental to deterrence. Granted, the indeterminate draw fails to reveal a clear path to defending important space capabilities, but neither does it indicate that potential adversaries have surefire ways to interdict them. Consequently, potential adversaries may well be dissuaded from developing many OSC

---

[3] This assumes, of course, that employing the DSC option would not require attacking Red's OSC capability preemptively. A preemptive attack would place the onus of escalation on the United States.

capabilities or deterred from employing those already available, given expectations of only mediocre mission performance and the probability of U.S. retribution. Likewise, U.S. leaders might also be dissuaded from developing OSC capabilities that offer only mediocre performance, considering the escalation risks and political costs associated with them. As a result, the indeterminate draw could be the outcome that offers the most first-strike stability in space, a condition that works to U.S. advantage.[4] U.S. developers can strengthen these deterrence dynamics by finding ways to raise potential attackers' political costs should they elect to develop or employ OSC capabilities despite their limited effectiveness.[5]

However, some OSC options that tend to appear in indeterminate-draws solution sets will not be stabilizing. In fact, some will likely be attractive to potential attackers even when other options offer higher mission effectiveness. These OSC options are those that are very cheap to develop, deploy, and employ, and particularly those that are difficult to attribute to the actor employing them. Certain reversible-effects attacks, such as terrestrial-based jamming, fit this description, and cyber attacks will be especially difficult to deter. Such capabilities are relatively inexpensive, so potential adversaries are likely to develop them even if their expected effectiveness is uncertain or doubtful. Whether potential adversaries actually employ such capabilities in a crisis or war will depend on what benefits they expect to achieve with them when compared with the risks and costs they generate. Adversaries will likely target U.S. systems with terrestrial-based jammers whether they are vulnerable to jamming or not. The costs of doing so are low, and U.S. leaders will have difficulty fashioning credible threats of retribution for non-kinetic, reversible-effects attacks. Similarly, opponents will attack with cyber weapons to the extent that the benefits they expect to gain by doing so outweigh the costs they would likely pay in lost access to targeted systems for purposes of intelligence exploitation. The

---

[4] One could argue that the United States and Soviet Union found themselves in an indeterminate draw in space throughout most of the Cold War, where both sides experimented with and could have fielded certain OSC systems, but few of the options available to them offered enough mission effectiveness to justify the escalation risks and political costs of employing them. Exceptions to this included Programs 437 and 505, the U.S. nuclear-armed ASAT systems fielded in the mid-1960s. These systems were developed despite the immense escalatory risks and political costs their employment would have entailed because of the highly threatening and escalatory system they were designed to engage, the nuclear-armed Fractional Orbital Bombardment System (FOBS) that Moscow was threatening to develop. For the history of the FOBS scare and the U.S. response to it, see Paul B. Stares, *The Militarization of Space: U.S. Policy 1945–1984*, Ithaca, N.Y.: Cornell University Press, 1985, pp. 59, 71, and 117–128; Clayton K. S. Chun, *Shooting Down a "Star": Program 437, the US Nuclear ASAT System and Present-Day Copycat Killers*, CADRE Paper No. 6, Maxwell AFB, Ala.: Air University Press, April 2000.

[5] For more on the concept of first-strike stability in space and ways to strengthen it, see Morgan, 2010, pp. 1–6 and 37–49.

best U.S. leaders will probably be able to do against these types of threats is identify them and develop defenses.[6]

Finally, even if the indeterminate draw does prove to be stabilizing for more-serious attacks, U.S. analysts should not become complacent in it; they should seek a path to a dominant strategy, knowing that potential adversaries are doing so as well. One possible approach would be to evaluate several of the top-performing offenses or defenses in combination to see if their strengths are mutually reinforcing or the strengths of one mitigates the weaknesses of another. Depending on such systems' costs, U.S. decisionmakers might choose to develop and deploy them in combination to better deter or defeat enemy attacks. In any event, despite the indeterminate nature of the draw, game-theoretic analysis will illuminate the strengths and weaknesses of each option, thereby revealing a developmental path to breaking the stalemate. In this respect, the indeterminate draw is potentially the most fruitful of all possible outcomes.

---

[6] For more on deterrence and escalation dynamics in cyberspace, see Martin C. Libicki, *Cyberdeterrence and Cyberwar*, Santa Monica, Calif.: RAND Corporation, MG-877-AF, 2009; Amir Lupovici, "Cyber Warfare and Deterrence: Trends and Challenges in Research," *Military and Strategic Affairs*, Vol. 3., No. 3, December 2011; Jonathan Soloman, "Cyberdeterrence Between Nation-States: Plausible Strategy or a Pipe Dream?" *Strategic Studies Quarterly*, Spring 2011; Martin C. Libicki, *Crisis and Escalation in Cyberspace*, Santa Monica, Calif.: RAND Corporation, MG-1215-AF, 2012.

6. Observations, Applications, and Recommendations

Observations

This report introduced a methodology based on sequential game theory that can be used to explore and better understand the potential dynamics of complex offense-defense interactions involving space capabilities. It also described a tool we have developed, the DSPAT, and another tool that could be developed, the MGT, to support analyses done within the game-theoretic framework. The DSPAT is versatile in its range of applications, and the MGT would add even more potential for expansion and modification. This methodology and the tools supporting it are valuable in that they offer a well-defined framework in which analysts can structure their assessments and solicit input from SMEs in a systematic manner.

The report described four types of analysis in which the game-theoretic methodology can be applied: a Dynamic Strategic Assessment evaluating all possible OSC and DSC matchups to identify the probable paths of conflict in space along with their associated levels of escalation and political cost; a Red Threat-Driven Assessment to identify the best defenses against specified space control threats; a Blue Technology-Driven Assessment to evaluate the mission effectiveness and potential deterrence and escalation implications of specific U.S. DSC program initiatives; and a Blue OSC Assessment to identify the most effective systems the United States could develop for interdicting opponent space capabilities with acceptable levels of escalation risk and political cost.

Finally, the report explained how to interpret the outcomes of analyses using the game-theoretic methodology. It described and discussed the implications of assessments that identify dominant strategies and saddle points, as well as those that end in indeterminate draws.

This chapter discusses potential applications for the game-theoretic methodology and its associated tools and offers recommendations to the U.S. Air Force.

Applications and Recommendations

The game-theoretic methodology provides a broad framework for assessing the potential effects of alternative OSC and DSC systems. The breadth and versatility of this analytical framework suggest a wide range of potential applications. We discuss several potential applications here and offer recommendations for their implementation.

Support for Program Development Decisions

Given that the inspiration for developing the game-theoretic methodology was to provide a means to assess the potential deterrence benefits and escalation risks of alternative DSC systems, program development decision support is its most obvious and direct application. The game-theoretic methodology provides an objective method of scoring and comparing alternative systems for their escalation risk and political cost, two areas that, because they defy precise quantification and measurement, are often undervalued or even ignored in program development decisions.

But the methodology does more than that. It also provides a structured framework for comparing approaches for their feasibility and mission effectiveness. In doing so, it directs attention to comprehensive lists of conceptual approaches to accomplishing whichever mission task is being examined, thereby widening the decision space to possibilities that might otherwise be overlooked. Once the most attractive concept is identified (and less effective or more dangerous alternatives discarded), the methodology provides an analytical foundation for program advocacy in a structured framework that can be easily tailored for briefings and papers.

Recommendations

1. Make the DSPAT available to selected system program offices and other agencies involved in national security space system research and development, and encourage them to learn and apply the game-theoretic methodology.
2. Insist upon thorough, method-based analysis in briefings and papers advocating the development of any OSC or DSC system.

Support for Operational Training and Tactical Decisionmaking

Although the game-theoretic methodology and DSPAT were devised for program development decision support, they can also make important contributions to ongoing national security space operations. The DSPAT, used independently or in conjunction with an MGT, can provide operational space commanders valuable analytical support for interpreting the implications of intelligence assessments, anticipating opposing space force behavior during crises, and developing the most effective and appropriate courses of action when threats to U.S. space capabilities manifest. The MGT, if developed as a comprehensive survey of attacks and defenses, would also offer operators an important resource for operational training in space warfare.

Recommendations

1. Make the DSPAT available to the Joint Space Operations Center (JSpOC) and selected Air Force space operations centers.

2. Incorporate use of the game-theoretic methodology in the JSpOC Mission System (JMS) and other operational decisionmaking tools and procedures there and at Air Force space operations centers.
3. Develop a comprehensive MGT and incorporate information from it in training programs at Air Force space operations units and centers, and train crew commanders on the tactical employment of method-based analysis.

Support for Space Play in War Games

Given the ever-increasing importance of space support to terrestrial forces in the emerging geostrategic environment, Service-, Joint-, and Office of the Secretary of Defense (OSD)-level war games often include Red attacks on U.S. national security space capabilities as important features of the scenarios played. These exercise events are intended to help U.S. warfighters think through the implications of such attacks and encourage them to develop appropriate responses. However, too often the inputs appear ad hoc without clear linkages to effects and are not clearly related to the broader Red concept of operations. On receiving such inputs, Blue teams often have difficulty responding to them in ways that balance the need to preserve space mission capability with the need to mitigate escalation risks and political costs. Finally, the adjudication of Red and Blue space play is mostly subjective and often not clearly related to effects on terrestrial operations.

The game-theoretic methodology can bring a higher level of coherence and objectivity to space play in war games. Game developers can use it to assess what Red OSC attacks would best support the scenario being designed and anticipate what would be good and bad Blue team responses to those attacks. Players on both sides could use the DSPAT to compare the escalation risks and political costs of space control options before deciding on appropriate courses of action. Adjudicators could assess Red and Blue inputs against each other in DSPAT and estimate space conflict outcomes, along with associated levels of escalation and political cost, with higher levels of analytical rigor than is possible with purely subjective assessments.

Recommendations

1. Make the game-theoretic methodology and any supporting tools available to Air Force, Joint, and OSD war game developers.
2. Encourage war game developers and controllers to make the DSPAT available to players.
3. Use the game-theoretic methodology and any supporting tools to support space event adjudication in major war games.

Support for Strategic Planning and Policymaking

One of the most important features of the game-theoretic methodology is its ability to do Dynamic Strategic Assessments of what OSC and DSC actions each belligerent would likely take at each level of confrontation and conflict in specified scenarios. These assessments of probable paths of conflict in space can apprise analysts, strategic planners, and policymakers on what levels of escalation to expect in those scenarios and what political costs leaders on both sides will incur if those expectations manifest. Such insights can inform force structure decisions (e.g., whether to focus more on developing OSC or DSC systems), doctrine and strategy development, and policy development for maximizing strategic stability and protecting national interests.

Recommendations

1. Make the game-theoretic methodology and supporting tools available to Air Force and Joint planners, space policymakers in OSD, and planning and policymaking offices in other national security space-related departments and agencies.
2. Conduct systematic, comprehensive Dynamic Strategic Assessments to determine likely paths of conflict in space in all defense planning scenarios and other scenarios of interest as they arise.

Appendix. Modeling the Assessment

In Chapter Three, we briefly described the DSPAT and how it is used in the game-theoretic methodology. This appendix explains how the DSPAT automates selected aspects of the scoring. It is not intended to serve as a DSPAT users' guide. DSPAT has more application modes, input menus, and output displays than are shown here. Rather, this appendix provides the central algorithms and tables that DSPAT uses to score OSC and DSC options and describes their user interfaces with illustrations of selected input menus.

Figure A.1. DSPAT OSC Criteria Menu

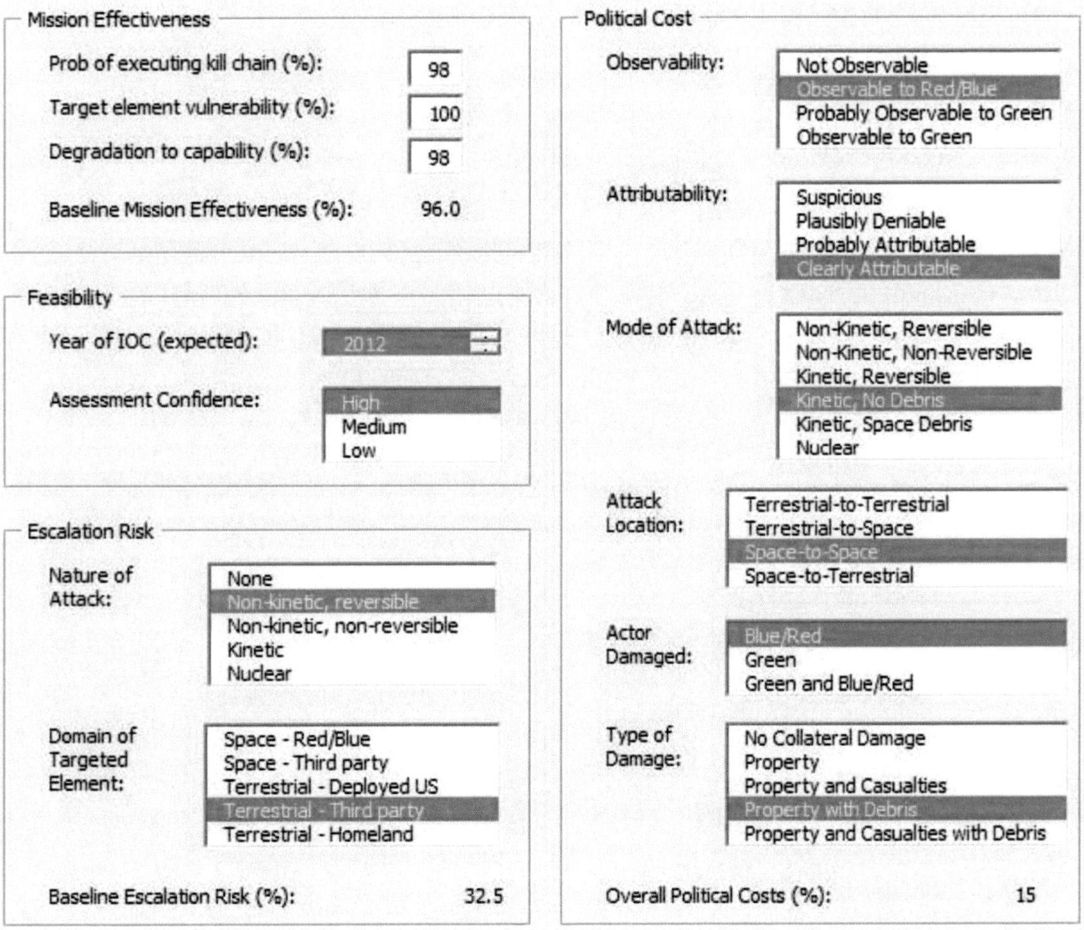

Scoring OSC Options

Any comparative assessment of multiple OSC options begins with a list of possible attacks on a selected space force enhancement capability. This list can be taken from an MGT, if one has been developed, or assembled separately by SMEs. To perform an assessment, analysts enter OSC options from the list into DSPAT and then use qualitative information provided by SMEs, or taken from an MGT if one is available, to help them fill out the DSPAT OSC Criteria Menu shown above in Figure A.1.

DSPAT solicits analysts' expert judgment to enter mission effectiveness and feasibility assessments and select the escalation risk and political cost parameters that best describe the nature of each potential attack. DSPAT scores each OSC option as it is entered, then ranks them all in each dimension (on a separate panel, DSPAT also asks analysts to enter time-to-effect assessments for each OSC option). DSPAT scores each assessment category using the following mathematical algorithms and tables.

Scoring Mission Effectiveness

As explained in Chapter Three, OSC mission effectiveness (ME) is the expected degradation to the targeted space capability resulting from one attack. Analysts are asked to score each component of mission effectiveness as a percentage from 100 to 0, with a score of 100 indicating a maximum level of mission effectiveness and a score of 0 indicating no mission effectiveness. Determining what values to enter in the mission effectiveness boxes will likely require assessment from both technical and operational SMEs who may be supported by computational resources separate from DSPAT.

DSPAT converts the percentages entered in the menu to probabilities (1.00 to 0.00) and calculates mission effectiveness as the product of three factors:

- the probability of a successful engagement, which is an assessment of the degree to which the attacker can execute each step of the kill chain in sequence (find, fix, track, target, and engage): PrSE
- expected degradation to the targeted element given a successful engagement: γ
- expected degradation to targeted capability given loss of the targeted element, which measures how much force enhancement capability is concentrated in, and dependent on, the targeted element: α.

Therefore,

$$\mathrm{ME} = \mathrm{PrSE} \times \gamma \times \alpha \qquad 0.00 \leq \mathrm{ME} \leq 1.00.$$

Scoring Feasibility

DSPAT treats the feasibility assessment as a filter. In a panel prior to the one shown in Figure A.1, it asks analysts to set a minimum threshold of feasibility acceptable for consideration in the assessment. Then, when assessing the feasibility of any particular OSC or DSC option, analysts enter the year for which the assessment is being conducted as shown in Figure A.1. Next, for each approach being evaluated, they enter an assessment of how confident they are, based on intelligence assessments for Red OSC options or system research and development assessments for Blue OSC options, that the option in question will reach initial operational capability (IOC) by that year. DSPAT retains in the solution set those options for which the feasibility is assessed as equal to or above the minimum threshold set in the model and discards those assessed as below that threshold.

Scoring Escalation Risk

The Escalation Risk section of the DSPAT OSC Criteria Menu (Figure A.1) solicits analysts to describe the nature of the OSC attack and the domain of the targeted element. It then uses those inputs to plot a location in the Quantitative Escalation Risk Matrix shown as Table A.1, which approximates the levels of vertical and horizontal escalation that the OSC attack would entail. DSPAT then draws a cell score from Table A.1, ranging from 0.00, indicating no escalation risk, to 1.00, indicating high escalation risk, which it shows as a baseline escalation score in turn one of the analysis and uses to calculate a conditional escalation score in subsequent turns.[1]

Table A.1. Quantitative Escalation Risk Matrix

Security	Space: RED/BLUE	Space: Third Party	Terrestrial: Forward-Deployed Forces	Terrestrial: Third Party Forces & Infrastructure	Terrestrial: RED/BLUE Homeland
Nuclear	0.875	0.890	0.950	0.975	1.000
Kinetic	0.625	0.750	0.750	0.775	0.800
Non-kinetic, non-reversible	0.500	0.625	0.500	0.563	0.625
Non-kinetic, reversible	0.375	0.500	0.125	0.250	0.375
No action / passive	0	0	0	0	0

Target Location

Key
0.950 to 1.000
0.800 to 0.950
0.700 to 0.800
0.575 to 0.700
0.425 to 0.575
0.300 to 0.425
0.200 to 0.300
0.050 to 0.200
0 to 0.050

[1] We shall explain the calculation of conditional escalation risk scores in the section entitled "Scoring DSC Options" later in this appendix.

Scoring Political Cost

As explained in Chapter Three, the space control game methodology defines *political costs* as those imposed on Red or Blue leaders by third-party observers, domestic and international. The level of political cost each OSC option generates is based on the degree to which third parties can observe and attribute it to the attacker, what parties are harmed or threatened by the attack, the degree of harm, and the degree to which the act violates international taboos on space warfare or the use of nuclear weapons.

DSPAT solicits analysts to select entries in the Political Cost section of the OSC Criteria Menu (Figure A.1) regarding details of each OSC option, which it converts to the following values:

Observability:

- Not Observable: 0.00
- Observable to Red/Blue: 0.50
- Probably Observable to Green: 0.75
- Observable to Green: 1.00

Attributability

- Suspicious 0.25
- Plausibly Deniable 0.50
- Probably Attributable 0.75
- Clearly Attributable 1.00

Nuclear and Space Warfare Taboo[2]

- Terrestrial to Terrestrial
  - Non-kinetic 0.00
  - Kinetic 0.00
  - Nuclear 1.00
- Terrestrial to Space
  - Non-Kinetic 0.25
  - Kinetic 0.50
  - Nuclear 0.90
- Space to Space
  - Non-Kinetic 0.35
  - Kinetic 0.60
  - Nuclear 0.95
- Space to Terrestrial
  - Non-Kinetic 0.45
  - Kinetic 0.75
  - Nuclear 1.00

---

[2] DSPAT identifies these levels based on entries made in the Mode of Attack and Attack Location boxes.

Collateral Damage

DSPAT selects entries made in the Mode of Attack, Attack Location, Actor Damaged, and Type of Damage boxes and converts the information to values drawn from Table A.2.

Table A.2. Collateral Damage Matrix

Damage to Non-Combatants	Blue/Red Terrestrial Property	Blue/Red Terrestrial Casualties	Blue/Red Space Property	Blue/Red Space Casualties	Green Terrestrial Property	Green Terrestrial Casualties	Green Space Property	Green Space Casualties	Green and Blue/Red Terrestrial Property	Green and Blue/Red Terrestrial Casualties	Green and Blue/Red Space Property	Green and Blue/Red Space Casualties
Non-Kinetic												
Reversible	0.05	0.25	0.15	0.35	0.35	0.50	0.20	0.60	0.40	0.55	0.25	0.65
Non-Reversible	0.15	0.35	0.25	0.45	0.45	0.60	0.50	0.70	0.50	0.65	0.55	0.75
Kinetic												
Reversible			0.25	0.45			0.35	0.70			0.40	0.75
Non-Reversible	0.15	0.35	0.25	0.45	0.45	0.60	0.55	0.70	0.50	0.65	0.60	0.75
With Space Debris			0.50	0.60			0.65	0.75			0.70	0.8
Nuclear	0.90	0.95	0.90	0.95	0.95	1.00	0.95	1.00	0.95	1.00	0.95	1.00

When all entries are complete, DSPAT calculates the political cost (PC) of the OSC option being examined as a relationship between the following factors:

- The degree to which the action is observable: s_1
- The degree to which the action is attributable to the party that committed it: s_2
- The degree to which the action violates a taboo: s_3
- The extent, location, and severity of collateral damage the action causes: s_4
- The respective weights assigned s_3 and s_4: w_3 and w_4 ($w_3 + w_4 = 1$)[1]

so that

$$PC = s_1 \times s_2 \times (w_3 \times s_3 + w_4 \times s_4) \qquad 0.00 \leq PC \leq 1.00.$$

A PC score of 0.00 indicates no political cost is accrued for the option being evaluated, while a PC score of 1.00 indicates substantial political costs. Unlike the way DSPAT treats escalation risk calculations, it does not condition political costs on whether attacks of comparable severity have been carried out in previous turns; each new attack generates political costs for the perpetrator derived from the values and formula above, regardless of whether similar or worse attacks have already taken place.

Scoring DSC Options

Once all OSC options are evaluated and the most likely option to be developed or employed is identified, DSPAT takes analysts to the panel shown in Figure A.2, DSPAT DSC Mission Effectiveness Menu, to begin scoring DSC options. At this point analysts should refer back to the MGT, if one is available, to identify all potential approaches for defending the threatened space capability from the specified attack. If no MGT is available, they could, with SME support, develop the list of DSC options manually. Analysts then enter DSC options from the list into DSPAT and, for each of them, complete the input panels as described in the following sections.

---

[1] The default settings for w_3 and w_4 in DSPAT are 0.50 and 0.50. These weights can be adjusted if subsequent research suggests one of these factors should be weighted heavier than the other.

Figure A.2. DSPAT DSC Mission Effectiveness Menu

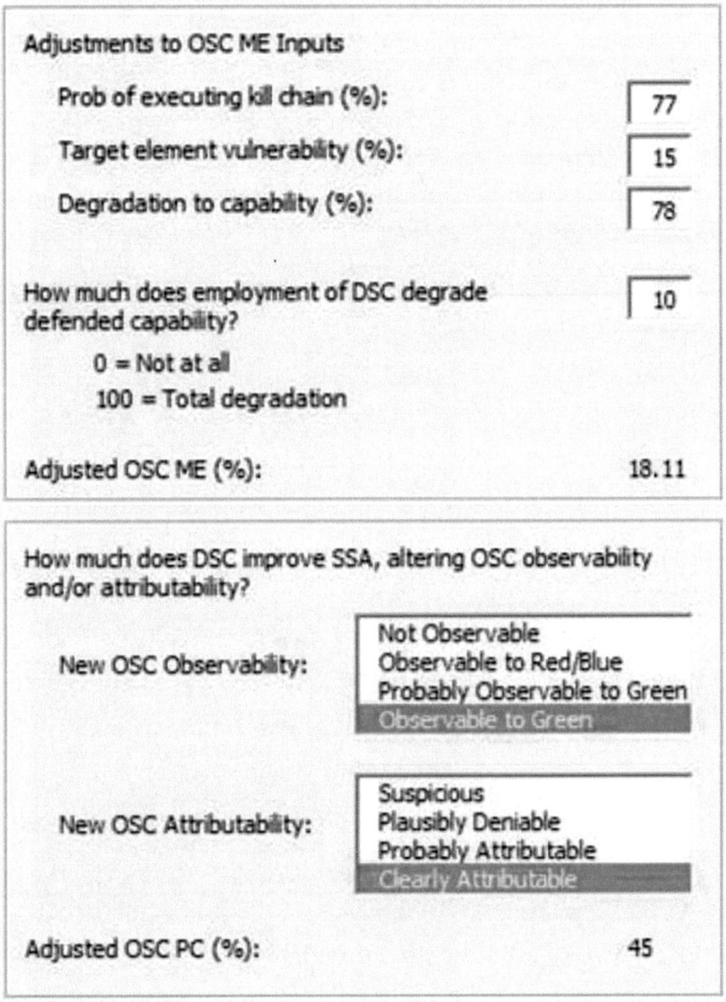

Scoring DSC Mission Effectiveness

As the figure indicates, some of the DSC menu items mirror those presented in the OSC menu in Figure A.1. However, it is important to remember that DSC mission effectiveness is not an independent calculation; rather, it is evaluated for how well any given option reduces the effectiveness of the OSC option it is meant to defend against. For a DSC option to have some positive effect, it must degrade one or more of the three factors used to compute OSC ME. Therefore, analysts are directed to enter percentages reflecting adjusted OSC ME factor scores that would result from employing the DSC. As in the OSC ME calculation, DSPAT converts the percentages entered into the menu to probabilities (1.00 to 0.00) and calculates an adjusted OSC ME as the product of three factors:

- adjusted probability of a successful OSC engagement, which is an assessment of the degree to which the attacker can still execute each step of the kill chain after the DSC is employed: <PrSE>
- adjusted expected degradation to the targeted element given a successful OSC engagement when the DSC is employed: <γ>
- adjusted expected degradation to targeted capability given loss of the target element, which indicates the degree to which employing the DSC changes how much force enhancement capability is concentration in, and dependent on, the targeted element: <α>.

Therefore, the adjusted OSC ME is

$$\langle \text{OSC ME} \rangle = \langle \text{PrSE} \rangle \times \langle \gamma \rangle \times \langle \alpha \rangle.$$

In most cases, a DSC option will degrade only one of the foregoing three factors. For instance, developing the capability for a satellite to detect an imminent kinetic attack and maneuver away from it might reduce a potential attacker's ability to successfully execute the kill chain for a kinetic OSC option. Analysts then might assess that, whereas the OSC PrSE was originally 0.75, the adjusted probability of successful engagement, <PrSE>, would be 0.50 using that DSC option. However, such a defense would not reduce the attack's effect on the targeted satellite should the kinetic strike succeed (i.e., <γ> would remain equal to γ). Nor would it reduce the degradation to the overall capability if the targeted element were lost (<α> would remain equal to α). Alternatively, developing a jam-resistant receiver for a communication satellite would not reduce an attacker's ability to execute the kill chain in a jamming option (<PrSE> would remain equal to PrSE), nor would it reduce degradation to the overall communication capability should the targeted element be lost (<α> would remain equal to α). However, it would reduce the expected effectiveness of the attack on that targeted element by a calculable amount—whereas γ might have been 0.90, <γ> might be, say, 0.30. Finally, a DSC option that adds robustness to the system, e.g., adding additional communications satellites or additional bandwidth on existing satellites, would not affect the attacker's ability to execute the kill chain or reduce the effect of a successful attack on any one targeted element. However, the extra capacity would reduce the overall degradation in capability if any one element were lost. Whereas the α for the successful destruction of one communication satellite in a four-ball constellation supporting a given theater would be 0.25, adding a fifth communications satellite would make <α> 0.20.[2] In any case, the effectiveness of a DSC

[2] This admittedly simplistic example assumes that the five satellites carry the same amount of bandwidth and all are fully subscribed to supporting the military mission. In actual assessments, analysts would need

option is the degree to which the adjusted OSC ME (<OSC ME>) is less than the initial OSC ME. In that regard, DSPAT treats DSC ME as the degree to which it protects the targeted space capability from a specific OSC attack. This relationship can be expressed as:

$$\text{DSC ME} = 1 - \langle \text{OSC ME} \rangle.$$

In addition to soliciting inputs for measuring the DSC option's effects on OSC ME, DSPAT asks analysts for three additional assessments: How much employing this DSC option would degrade the defended mission capability, how much it would make the OSC option it is defending against more observable to Green, and how much it would make the OSC option more attributable to the attacker. The answer to the first question directly affects DSC ME as a relationship between the following factors:

- Adjusted DSC ME: <DSC ME>
- Degradation to defended force enhancement capability: DgFECap

so that

$$\langle \text{DSC ME} \rangle = \text{DgFECap} \times \text{DSC ME} \qquad 0.00 \leq \langle \text{DSC ME} \rangle \leq 1.00.$$

If <DSC ME> equals 1.00, then employment of the Blue DSC option would enable the threatened Blue space capability to maintain full operational capability under attack by the Red OSC option; on the other hand, a <DSC ME> score of 0.00 indicates a complete loss of Blue space system utility.

Answers to the second and third questions adjust the attacker's political cost score by raising the s_1 and s_2 values associated with that specific OSC attack (see the section "Scoring Political Cost" earlier in this appendix).

Scoring DSC Feasibility, Escalation Risk, and Political Cost

After DSC mission effectiveness inputs are entered, DSPAT takes analysts to the panel shown in Figure A.3, DSPAT DSC feasibility, escalation risk, and political cost menu.

---

to survey or at least estimate the amount of bandwidth available on each satellite and calculate α and $\langle\alpha\rangle$ accordingly.

Figure A.3. DSPAT DSC Feasibility, Escalation Risk, and Political Cost Menu

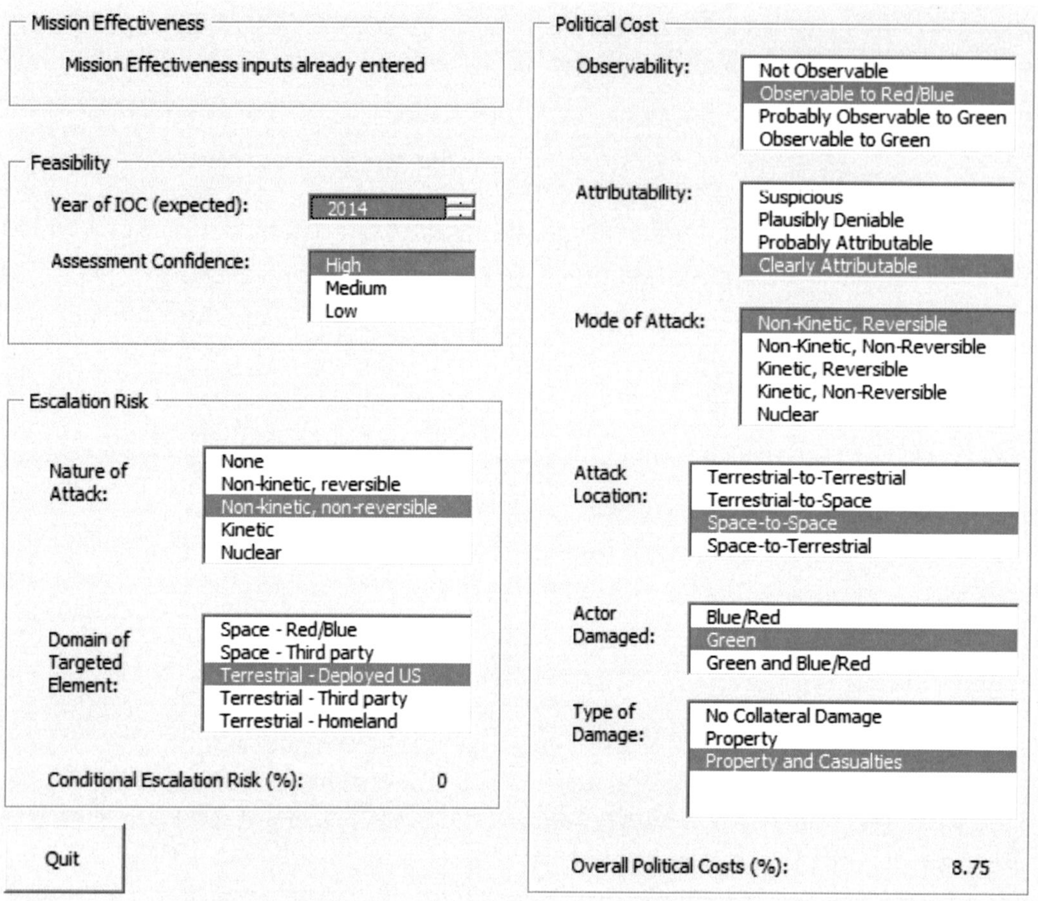

DSPAT treats DSC feasibility the same way it handles that assessment for OSC options. DSC feasibility is an estimate of whether the option under consideration will reach IOC by a specified year. As is the case with OSC options, those estimated to be available with high or medium confidence are kept in the solution set; those not expected to be available, or expected with only low confidence, are discarded. Analysts should consult system research and development sources for projections on Blue DSC option availability and intelligence assessments for Red DSC option availability.

Escalation risk scoring for DSC options begins with an approach mirroring that is used in calculating escalation risks for OSC options. The Escalation Risk section of the DSPAT DSC feasibility, escalation risk, and political cost menu solicits inputs regarding the nature of the attack and the domain of the targeted element. Use of the word "attack" acknowledges that some active defenses involve attacking the OSC system that is threatening the defended system. To score passive defenses and options that do not involve attacking, analysts select "None" in the Nature of Attack box. Based on these inputs, DSPAT plots the option on Table A.1, Quantitative Escalation Risk Matrix, and draws a cell score from 0.00, indicating no escalation risk, to 1.00, indicating high escalation risk.

Because most DSC options that involve active defenses are initiated in response to an OSC attack, the DSC escalation risk is conditional and depends on the level of escalation already generated in the OSC attack. All OSC and DSC options in subsequent turns also require conditional escalation calculations. For each OSC or DSC action taken after the first turn of a game-theoretic analysis, DSPAT divides its cell score by the baseline score to obtain the conditional escalation risk (but keeping the value ≤ 1), as indicated:

$$ER_{CondDSC} = \begin{cases} 1 - \dfrac{1 - ER_{DSC}}{1 - EROSC}, & if\ ER_{OSC} < ER_{DSC}\ ,\ ER_{OSC} < 1. \\ 0 \quad , & otherwise \end{cases}$$

Some DSC options, however, entail launching preemptive attacks. In those cases, DSC escalation risks are calculated in the same manner as they are for first-turn OSC options, and they set baseline escalation scores.

Finally, DSPAT calculates political costs for DSC options using the same algorithm and table used for calculating political costs for OSC options (see "Scoring Political Cost" section and Table A.2 in this appendix). As in OSC scoring, DSPAT does not condition political costs of DSC actions on whether attacks of comparable severity have been carried out in previous turns; each new act generates political costs for the perpetrator derived from the values and formula above, regardless of whether similar or worse attacks have already taken place.

References

AFSPC—*See* Air Force Space Command.

Air Force Space Command, "Space and Cyberspace Priorities," undated, publicized July 16, 2012. As of September 4, 2013: http://www.airforcemag.com/SiteCollectionDocuments/Reports/2012/July%202012/Day17/071612_AFSPC_space_cyber_priorities.pdf

"Air Force Space Command Defines Its Priorities," *Air Force Magazine,* July 17, 2012. As of September 4, 2013: http://www.airforcemag.com/DRArchive/Pages/2012/July%202012/July%2017%202012/AirForceSpaceCommandDefinesitsPriorities.aspx

Albert, Michael H., and Richard J. Nowakowski, eds., *Games of No Chance 3*, Vol. 56, Mathematical Sciences Research Institute Publications, Cambridge, UK: Cambridge University Press, 2009.

Bao Shixiu, "Deterrence Revisited: Outer Space," *China Security*, No. 5, Winter 2007. As of July 21, 2017: http://www.issuelab.org/resources/446/446.pdf

Berlekamp, Elwyn, John H. Conway, and Richard Guy, *Winning Ways*, Vols. 1–2, London: Academic Press, 1982.

"China Confirms Satellite Downed," *BBC News,* January 23, 2007. As of September 4, 2013: http://news.bbc.co.uk/2/hi/asia-pacific/6289519.stm

Chun, Clayton K. S., *Shooting Down a "Star": Program 437, the U.S. Nuclear ASAT System and Present-Day Copycat Killers*, CADRE Paper No. 6, Maxwell Air Force Base, Ala.: Air University Press, April 2000.

Conway, John H., *On Numbers and Games*, London: Academic Press, 1976.

Council of the European Union, "Code of Conduct for Outer Space Activities," revised draft, Brussels, October 11, 2010. As of September 9, 2013: http://www.consilium.europa.eu/uedocs/cmsUpload/st14455.en10.pdf

Deblois, Bruce M., "Space Sanctuary: A Viable National Strategy," *Airpower Journal,* Winter 1998.

Deblois, Bruce M., ed., *Beyond the Paths of Heaven: The Emergence of Space Power Thought*, Maxwell Air Force Base, Ala.: Air University Press, 2003.

Demaine, Eric D., and Robert A. Hearn, "Playing Games with Algorithms: Algorithmic Combinatorial Game Theory," in Michael H. Albert and Richard J. Nowakowski, eds., *Games of No Chance 3*, Vol. 56, Mathematical Sciences Research Institute Publications, Cambridge, UK: Cambridge University Press: 2009, pp. 3–56.

Department of Defense Directive 3100.10, *Space Policy,* Washington, D.C., U.S. Department of Defense, October 18, 2012, incorporating Change 1, effective November 4, 2016. As of July 21, 2017: https://www.hsdl.org/?view&did=796643

Dixit, Avinash K., and Barry J. Nalebuff, *Thinking Strategically: The Competitive Edge in Business, Politics, and Everyday Life*, New York: W. W. Norton & Co., 1991.

Dolman, Everett, *Astropolitik: Classical Geopolitics in the Space Age*, London: Frank Cass, 2002.

Finch, James P., and Shawn Steene, "Finding Space in Deterrence: Toward a General Framework for 'Space Deterrence,'" *Strategic Studies Quarterly*, Winter 2011.

Gopalaswamy, Bhrarath, and Harsh V. Pant, "Does India Need Anti-Satellite Capability?" *Rediff News*, February 9, 2010. As of September 4, 2013: http://www.rediff.com/news/column/does-india-need-anti-satellite-capability/20100209.htm

Gray, Colin, "Space Arms Control: A Skeptical View," *Air University* Review, November–December 1985.

Knuth, Donald E., *Surreal Numbers*, Reading, Mass.: Addison-Wesley, 1974.

Krepon, Michael, Theresa Hitchens, and Michael Katz-Hyman, "Preserving Freedom of Action in Space: Realizing the Potential and Limits of U.S. Spacepower," in Charles D. Lutes and Peter L. Hays, eds., *Toward a Theory of Spacepower: Selected Essays*, Washington, D.C.: National Defense University Press, 2012, pp. 119–136.

Libicki, Martin C., *Cyberdeterrence and Cyberwar*, Santa Monica, Calif.: RAND Corporation, MG-877-AF, 2009. As of September 9, 2013: http://www.rand.org/pubs/monographs/MG877.html

———, *Crisis and Escalation in Cyberspace*, Santa Monica, Calif.: RAND Corporation, MG-1215-AF, 2012. As of September 9, 2013: http://www.rand.org/pubs/monographs/MG1215.html

Lupovici, Amir, "Cyber Warfare and Deterrence: Trends and Challenges in Research," *Military and Strategic Affairs*, Vol. 3, No. 3, December 2011.

Lutes, Charles D., and Peter L. Hays, eds., *Toward a Theory of Spacepower: Selected Essays*, Washington, D.C.: National Defense University Press, 2012.

Miles, Donna, "Iraq Jamming Incident Underscores Lessons About Space," *American Forces Press Service*, September 15, 2004. As of April 5, 2017: http://archive.defense.gov/news/newsarticle.aspx?id=25298

Morgan, Forrest E., *Deterrence and First-Strike Stability in Space: A Preliminary Assessment*, Santa Monica, Calif.: RAND Corporation, MG-916-AF, 2010. As of September 9, 2013: http://www.rand.org/pubs/monographs/MG916.html

Morgan, Forrest E., Karl P. Mueller, Evan S. Medeiros, Kevin L. Pollpeter, and Roger Cliff, *Dangerous Thresholds: Managing Escalation in the 21st Century*. Santa Monica, Calif.: RAND Corporation, MG-614-AF, 2008. As of September 9, 2013: http://www.rand.org/pubs/monographs/MG614.html

National Science and Technology Council, "Fact Sheet: National Space Policy," White House, Office of Science and Technology Policy, Washington, D.C., September 19, 1996. As of May 5, 2013: https://ntrs.nasa.gov/archive/nasa/casi.ntrs.nasa.gov/20010011194.pdf

National Space Policy Directive-1, *National Space Policy Directives and Executive Charter,* Washington, D.C.: White House, November 2, 1989. As of September 4, 2013: http://www.au.af.mil/au/awc/awcgate/nspd1.htm

Reinsch, William A., and Dennis C. Shea, *2015 Annual Report to Congress of the U.S.-China Economic and Security Review Commission*, 114th Congress, First Session, Chapter 2, Section 2, Washington, D.C.: U.S. Government Printing Office, November 2015, p. 292.

"Satellite Test Sparks Overblown Worries," *Global Times,* January 6, 2013. As of September 4, 2013: http://www.globaltimes.cn/content/753925.shtml

Schulte, Ambassador Gregory L., Deputy Assistant Secretary of Defense for Space Policy, "Protecting Global Security in Space," presentation at the S. Rajaratnam School of International Studies, Nanyang Technological University, Singapore, May 9, 2012. As of September 4, 2013:

http://archive.defense.gov/home/features/2011/0111_nsss/docs/Rajaratnam%20School%20of%20International%20Studies%20on%20Protecting%20Global%20Security%20in%20Space,%20May%209,%202012.pdf

Soloman, Jonathan, "Cyberdeterrence Between Nation-States: Plausible Strategy or a Pipe Dream?" *Strategic Studies Quarterly*, Spring 2011.

Stares, Paul B., *The Militarization of Space: U.S. Policy 1945–1984*, Ithaca, N.Y.: Cornell University Press, 1985.

U.S. Department of Defense, "Resilience of Space Capabilities," fact sheet, Washington, D.C., 2011. As of July 21, 2017:
http://archive.defense.gov/home/features/2011/0111_nsss/docs/DoD%20Fact%20Sheet%20-%20Resilience.pdf

U.S. Department of Defense and Office of the Director of National Intelligence, *National Security Space Strategy: Unclassified Summary*, Washington, D.C., January 2011. As of September 4, 2013:
https://www.hsdl.org/?view&did=10828

White House, "Presidential Directive on National Space Policy," unclassified summary, Washington, D.C., February 11, 1988. As of April 5, 2017:
http://www.hq.nasa.gov/office/pao/History/policy88.html

———, "U.S. National Space Policy," Washington, D.C., August 31, 2006. As of September 9, 2013:
https://fas.org/irp/offdocs/nspd/space.pdf

———, *National Space Policy of the United States of America*, Washington, D.C., June 28, 2010. As of September 9, 2013:
https://obamawhitehouse.archives.gov/sites/default/files/national_space_policy_6-28-10.pdf

Williams, J. D., *The Compleat Strategyst: Being a Primer on the Theory of Games of Strategy*, New York: Dover Publications, Inc., 1986.

Zeigler, David W., "Safe Heavens: Military Strategy and Space Sanctuary," in Bruce M. Deblois, ed., *Beyond the Paths of Heaven: The Emergence of Space Power Thought*, Maxwell Air Force Base, Ala.: Air University Press, 2003, pp. 185–245.

Zhengyan Fang, "America's Space Fighter: New Threat to Humanity," *Ta Kung Pao*, May 5, 2010. As of September 9, 2013:
http://watchingamerica.com/News/56135/america%E2%80%99s-space-fighter%C2%A0new-threat-to-humanity